AF361903

Surface Treatment by
Laser-Assisted Techniques

Surface Treatment by Laser-Assisted Techniques

Editor

Rafael Comesaña

MDPI • Basel • Beijing • Wuhan • Barcelona • Belgrade • Manchester • Tokyo • Cluj • Tianjin

Editor
Rafael Comesaña
University of Vigo
Spain

Editorial Office
MDPI
St. Alban-Anlage 66
4052 Basel, Switzerland

This is a reprint of articles from the Special Issue published online in the open access journal *Coatings* (ISSN 2079-6412) (available at: https://www.mdpi.com/journal/coatings/special_issues/surface_laser).

For citation purposes, cite each article independently as indicated on the article page online and as indicated below:

LastName, A.A.; LastName, B.B.; LastName, C.C. Article Title. *Journal Name* **Year**, *Article Number*, Page Range.

ISBN 978-3-03943-184-7 (Hbk)
ISBN 978-3-03943-185-4 (PDF)

© 2020 by the authors. Articles in this book are Open Access and distributed under the Creative Commons Attribution (CC BY) license, which allows users to download, copy and build upon published articles, as long as the author and publisher are properly credited, which ensures maximum dissemination and a wider impact of our publications.

The book as a whole is distributed by MDPI under the terms and conditions of the Creative Commons license CC BY-NC-ND.

Contents

About the Editor . . . . **vii**

Rafael Comesaña
Special Issue on Surface Treatment by Laser-Assisted Techniques
Reprinted from: *Coatings* **2020**, *10*, 580, doi:10.3390/coatings10060580 . . . . **1**

**Mengyao Liu, Rui Zhou, Zhekun Chen, Huangping Yan, Jingqin Cui, Wanshan Liu,
Jia Hong Pan and Minghui Hong**
Tunable Hierarchical Nanostructures on Micro-Conical Arrays of Laser Textured TC4 Substrate
by Hydrothermal Treatment for Enhanced Anti-Icing Property
Reprinted from: *Coatings* **2020**, *10*, 450, doi:10.3390/coatings10050450 . . . . **7**

**Vanna Torrisi, Maria Censabella, Giovanni Piccitto, Giuseppe Compagnini,
Maria Grazia Grimaldi and Francesco Ruffino**
Characteristics of Pd and Pt Nanoparticles Produced by Nanosecond Laser Irradiations of Thin
Films Deposited on Topographically-Structured Transparent Conductive Oxides
Reprinted from: *Coatings* **2019**, *9*, 68, doi:10.3390/coatings9020068 . . . . **21**

**Mònica Fernández-Arias, Massimo Zimbone, Mohamed Boutinguiza, Jesús Del Val,
Antonio Riveiro, Vittorio Privitera, Maria G. Grimaldi and Juan Pou**
Synthesis and Deposition of Ag Nanoparticles by Combining Laser Ablation and
Electrophoretic Deposition Techniques
Reprinted from: *Coatings* **2019**, *9*, 571, doi:10.3390/coatings9090571 . . . . **39**

**Francesco Baino, Maria Angeles Montealegre, Joaquim Minguella-Canela
and Chiara Vitale-Brovarone**
Laser Surface Texturing of Alumina/Zirconia Composite Ceramics for Potential Use in Hip
Joint Prosthesis
Reprinted from: *Coatings* **2019**, *9*, 369, doi:10.3390/coatings9060369 . . . . **49**

**Adham Al-Sayyad, Julien Bardon, Pierre Hirchenhahn, Regis Vaudémont, Laurent Houssiau
and Peter Plapper**
Influence of Aluminum Laser Ablation on Interfacial Thermal Transfer and Joint Quality of
Laser Welded Aluminum–Polyamide Assemblies
Reprinted from: *Coatings* **2019**, *9*, 768, doi:10.3390/coatings9110768 . . . . **65**

Ismail M. Tayel
Thermoelastic Response Induced by Volumetric Absorption of Uniform Laser Radiation
in a Half-Space
Reprinted from: *Coatings* **2020**, *10*, 228, doi:10.3390/coatings10030228 . . . . **79**

Yongquan Zhou, Zhenyu Zhao, Wei Zhang, Haibing Xiao and Xiaomei Xu
Experiment Study of Rapid Laser Polishing of Freeform Steel Surface by Dual-Beam
Reprinted from: *Coatings* **2019**, *9*, 324, doi:10.3390/coatings9050324 . . . . **97**

**Joaquín Penide, Jesús del Val, Antonio Riveiro, Ramón Soto, Rafael Comesaña,
Félix Quintero, Mohamed Boutinguiza, Fernando Lusquiños and Juan Pou**
Laser Surface Blasting of Granite Stones Using a Laser Scanning System
Reprinted from: *Coatings* **2019**, *9*, 131, doi:10.3390/coatings9020131 . . . . **109**

Haijiang Wang, Wei Zhang, Yingbo Peng, Mingyang Zhang, Shuyu Liu and Yong Liu
Microstructures and Wear Resistance of FeCoCrNi-Mo High Entropy Alloy/Diamond
Composite Coatings by High Speed Laser Cladding
Reprinted from: *Coatings* **2020**, *10*, 300, doi:10.3390/coatings10030300 . . . . **121**

**Thomas Schopphoven, Norbert Pirch, Stefan Mann, Reinhart Poprawe,
Constantin Leon Häfner and Johannes Henrich Schleifenbaum**
Statistical/Numerical Model of the Powder-Gas Jet for Extreme High-Speed Laser
Material Deposition
Reprinted from: *Coatings* **2020**, *10*, 416, doi:10.3390/coatings10040416 . . . . **137**

Kaijin Huang, Wei Li, Kai Pan, Xin Lin and Aihua Wang
High Temperature Oxidation and Thermal Shock Properties of La$_2$Zr$_2$O$_7$ Thermal Barrier
Coatings Deposited on Nickel-Based Superalloy by Laser-Cladding
Reprinted from: *Coatings* **2020**, *10*, 370, doi:10.3390/coatings10040370 . . . . **155**

About the Editor

Rafael Comesaña received his M.SC in Mechanical Engineering from the University of Vigo, Spain, in 2006. In the LaserON—Laser Applications Research Group of the University of Vigo, he conducted his Ph.D. research on additive manufacturing based on laser cladding for the production of bone functional implants, graduating in 2010. He performed research stays in the Department of Materials of the Imperial College London, in the Laboratoire pour l'Application des Laser de Puissance—LALP of Arts et Métiers ParisTech, in the Barts, and The London School of Medicine and Dentistry, and in the Institute of Biomedical Optics of the University of Lübeck. In 2012, he joined the Department of Materials Engineering, Applied Mechanics, and Construction of the University of Vigo, where he is associated professor.

Editorial

Special Issue on Surface Treatment by Laser-Assisted Techniques

Rafael Comesaña [1,2]

[1] Department of Materials Engineering, Applied Mechanics and Construction, University of Vigo, EEI, Lagoas-Marcosende, 36310 Vigo, Spain; racomesana@uvigo.es
[2] LaserON Laser Applications Research Group, University of Vigo, Industrial Technological Research Centre - MTI, Rúa Maxwel, 36310 Vigo, Spain

Received: 3 June 2020; Accepted: 15 June 2020; Published: 22 June 2020

1. Introduction

Laser radiation is a powerful tool for surface modification owing to its spatial and material absorbance selectivity. The non-contact and chemically clean characteristics of laser processing make this technique very attractive for surface treatment in a wide variety of scientific and engineering areas. Moreover, the advances made in laser pulse control and beam delivery, together with increasingly available and affordable laser sources, has boosted laser integration in surface modification and coating procedures at research and industry levels.

The applications of enhanced surfaces by laser-assisted treatments are very broad, and only some of them are introduced here. The fabrication of superhydrophobic, self-cleaning, and anti-icing surfaces has a potential impact in the aircraft industry by improving the structural behavior at low temperatures. Surfaces for sensing, catalytic, electronic, biomedical, as well as energy production and storage purposes are modified at nanometer level to take advantage of their high specific surface area and reactivity. In the field of the biomedical implants, the enhancement of biomedical hip join prosthesis is addressed to improve surface fixation to host bone, avoiding the use of external acrylic cements. In the manufacturing industry, free-form surface molds can be efficiently polished by laser radiation-based systems. Lightweight hybrid structures with tailored properties can benefit from the production of metal-polymer dissimilar joints, where the structural properties are improved by the seam surface modification prior to thermal joining. Stone and mining industries require highly abrasive anti-wear components for cutting saws, which can be obtained by laser cladding of metal-diamond composites. Moreover, a stone cut surface finish can be tailored with minimal dust and no tool wear if the mechanical and flame equipment are substituted by the laser blasting technique. Finally, thermal barrier coatings for operation temperatures over 1000 °C for the aerospace, petrochemical, and energy industries can be obtained by laser cladding on metallic superalloys.

The most important laser techniques in surface engineering can be divided into the four main types of laser treatment: remelting-free techniques, remelting techniques, evaporation techniques and the special techniques for the formation of thin and hard coatings. Laser remelting-free techniques are represented, e.g., by surface annealing, tempering or preheating, surface hardening, and surface cleaning. Laser remelting techniques include surface remelting, surface alloying, and cladding. The most commonly used laser evaporation techniques are as follows: pure evaporation, detonation hardening and ablation. The technique referred as laser texturing usually involves material ablation and it can be classified as an evaporation technique, but occasionally mixed melting/vaporization are found below this term. The thin and hard coatings could be formed by the fusion alloying in the gas method, pure vapor deposition, pyrolytic, and photochemical methods as well as by chemical methods, e.g., laser-assisted chemical vapor deposition.

2. Surface Treatment by Laser-Assisted Techniques

The eleven research articles of this Special Issue "Surface Treatment by Laser-Assisted Techniques" cover laser surface modification from the nanoscale to the macroscale. Specific topics range from the production of nanoparticle-structured thin films to the deposition of mm-thick coatings, passing though the micro-texturing of metallic and ceramic surfaces. The individual work is summarised below:

The paper "Tunable Hierarchical Nanostructures on Micro-Conical Arrays of Laser Textured TC4 Substrate by Hydrothermal Treatment for Enhanced Anti-Icing Property" by Liu et al. presents the fabrication of anti-icing structured surfaces on Ti6Al4V alloy substrates [1]. Hydrothermal treatment in aqueous alkali of the microstructured laser-ablated surface led to the formation of titania (TiO_2) nanostructures. This work discusses the nanoarray growth mechanism on the micro-conical arrays and shows its superhydrophobic (approximately contact angle of 160°) and water freezing delaying properties.

The paper "Characteristics of Pd and Pt Nanoparticles Produced by Nanosecond Laser Irradiations of Thin Films Deposited on Topographically-Structured Transparent Conductive Oxides" authored by Torrisi et al. shows the production of Pd and Pt nanoparticles on Fluorine-doped tin oxide (FTO) by laser-assisted dewetting of nanoscale-thick films [2]. This work explains how the substrate topography plays a role in the liquid metal film dewetting and in the associated nanoparticle characteristics, and the existence of a critical film thickness related to the change of dewetting characteristics of the film. Moreover, a surface enhancement Raman scattering effect (SERS) was observed in the Pd nanoparticle/FTO/glass samples.

The paper "Synthesis and Deposition of Ag Nanoparticles by Combining Laser Ablation and Electrophoretic Deposition Techniques" by Fernández-Arias et al. presents the production of uniform silver nanostructured thin films on silicon substrate by pulsed laser ablation in liquid (PLAL) and electrophoretic deposition (ED) techniques [3]. During the ablation process in aqueous media, the silver target constitutes the positive electrode, while the silicon substrate is the negative electrode. This work addresses the analysis of the surface topography, composition, crystallinity and optical properties of the produced thin films, and localized the surface plasmon resonance (LSPR) around 400 nm.

The paper "Laser Surface Texturing of Alumina/Zirconia Composite Ceramics for Potential Use in Hip Joint Prosthesis" authored by Baino et al. describes the texturization of bioinert composite ceramics to improve fixation to host bone of hip joint acetabular cups [4]. The incidence of the laser processing parameters in the modified surface topography is methodically analysed in this work. The surface roughness was observed to be modified between in the range of 3–30 µm, and the application to real ceramic acetabular cups with a curved profile was demonstrated to be feasible.

The paper "Influence of Aluminum Laser Ablation on Interfacial Thermal Transfer and Joint Quality of Laser Welded Aluminum–Polyamide Assemblies" by Al-sayyad et al. presents the laser ablation of aluminium alloy for aluminium-polyamide 6.6 dissimilar joining and investigates the effects of the laser irradiation on the modified surface properties [5]. Laser flash analysis (LFA) and thermal contact resistance (TCR) quantification in the obtained surfaces is performed. A strong influence of laser ablation parameters on the surface structural and morphological properties is evidenced.

The paper "Thermoelastic Response Induced by Volumetric Absorption of Uniform Laser Radiation in a Half-Space" by Tayel implements the application of the generalized theory with Dual-Phase-Lag (DPL) to the study of the thermoelastic response for pulsed laser absorption [6]. Predictions in the practical case of copper media are analysed and compared to application of Lord-Shulman (LS) and classical coupled (CTE) theories. The results give insight to the expected evolution of several important parameters in pulsed laser surface modification, for instance, temperature lag, temperature distribution, and stress distribution.

The paper "Experiment Study of Rapid Laser Polishing of Freeform Steel Surface by Dual-Beam" authored by Zhou et al. focuses on the production of steel polished surfaces by combination of a top-hat continuous wave beam and top-hat pulsed laser beam [7]. The influence of the initial surface roughness, the laser spot size, the scanned trajectory, and the waveform of the pulsed laser on the polished surface topography is analysed. In this work, post-polished promising values down to 142 nm in average roughness were achieved, opening the door to a highly efficient final polishing step for free-form surfaces.

The paper "Laser Surface Blasting of Granite Stones Using a Laser Scanning System" by Penide et al. presents the surface texturization of granite stones by means of scanned laser irradiation in air atmosphere [8]. This work explains, through an experimental factorial design, the influence of the processing parameters on CO_2 laser roughening from as cut or polished state up to average roughness values of 20 μm. Preservation of quartz and feldspar crystallinity was observed, while annite experienced amorphization and micro-droplet formation. The authors demonstrate the induction of less mechanical stresses than in conventional bush hammering or flame blasting, therefore this technique can process lower thickness granite tiles.

The paper "Microstructures and Wear Resistance of FeCoCrNi-Mo High Entropy Alloy/Diamond Composite Coatings by High Speed Laser Cladding" by Wang et al. is devoted FeCoCRNi-Mo high entropy alloy/diamond composite coatings by laser cladding to be used as highly abrasive and wear resistance surface [9]. The authors explain the components proportion and processing parameters optimization to avoid graphitization and diamond thermal damage, and to improve bonding to the metallic matrix. In this work, the successful production of composite coatings with uniform microstructure and good wear resistance was demonstrated.

The paper "Statistical/Numerical Model of the Powder-Gas Jet for Extreme High-Speed Laser Material Deposition" authored by Schopphoven et al. describes the gas-powder stream modellization with practical application in laser cladding, and in additive manufacturing based on laser cladding (laser material deposition, LMD) [10]. This mathematical and experimental work considers coaxial powder delivery and spherical precursor material, and provides useful information for the prediction of precursor particle trajectories and particle-radiation interactions in the jet and laser beam intersection volume.

The paper "High Temperature Oxidation and Thermal Shock Properties of $La_2Zr_2O_7$ Thermal Barrier Coatings Deposited on Nickel-Based Superalloy by Laser-Cladding" by Huang et al. presents the production of thermal barrier coatings on a superalloy substrate and its characterization after isothermal oxidation and thermal cycling up to 1100 °C [11]. Preservation of the $La_2Zr_2O_7$ coating composition after processing was observed, and moderated improvements in the thermal shock lifetime and oxidation weight gain in comparison to the superalloy substrate were demonstrated.

3. Perspectives

In the laser-assisted techniques, the wavelength, pulse length, pulse energy, and associated surface irradiation density play a principal role to delimit the surface modification capabilities for a specific material. No single laser source is able to provide the optimal optical radiation for a specific surface treatment. Thus, having a clear overview of the laser beam capabilities is important for a correct research design. The applied laser treatment and optical radiation for each material surface in the scientific works collected by this special issue are recapitulated in the following paragraphs.

Laser texturing of titanium alloy is performed by means of a Nd:YAG pulsed laser with a wavelength of 1064 nm, 100 ns pulse width, and 0.1 mm spot size [1]. Laser ablation of Pd and Pt is achieved via nanosecond pulsed Nd:YAG laser working at 532 nm, 12 ns pulse length, and fluence of 0.50 J/cm^2 [2]. Laser ablation of Ag nanoparticles in liquid media is done by the use of of nanosecond pulsed Nd:YVO4 laser source operating at 532 nm, 14 ns pulse length, and average optical power of 6 W [3]. Laser texturing of ceramic composites is made by means of Q-switched Nd:YVO4 laser source operating at 1064 nm, 10 ns pulse length, and pulse energies in the range 240–800 μJ [4]. Laser ablation of aluminium alloy as pre-processing for laser assisted metal-polymer joining is accomplished by using a Nd:YVO4 laser source operating at 1064 nm, and fluence values in the range 8.7–28.6 J/cm^2, and subsequent laser joining by a fiber laser operating at 1070 nm, and 35 μs pulse length [5].

Laser polishing of steel is implemented via combination of two top-hat laser beams, a continuous wave fibre laser delivering 600 W at 1080 nm and 0.47 mm spot size, and a pulsed Q-switched fibre laser delivering 80 W at 1064 nm, pulse length 1.3 µs and 0.32 mm spot size [7]. Laser surface blasting of granite stone is performed by means 1 kW of CO_2 laser radiation operating at 10600 nm, a 710 mm focal length, a polygonal scanner, and 0.56 mm spot size [8].

Metal-diamond composite coatings are produced by laser cladding by the use of continuous wave diode laser irradiation with optical power in the range 3.0–5.5 kW, robot scanning speed between 30 and 60 mm/s, 4.6 mm spot size, and coaxial powder delivery [9]. Thermal barrier ceramic coatings on a high temperature nickel super alloy are deposited by laser cladding by means of continuous wave diode laser irradiation with optical power in the range 3.5 kW, 10 mm/s scanning speed, 5 mm × 5 mm spot size, and preplaced powder [11].

It can be outlined the use of the nanosecond pulsed laser radiation in the visible and near-infrared ranges of the spectra for micro- and nano-texturing of a wide range of materials, while the semiconductor diode laser radiation in the near-infrared range is employed for laser cladding. Moreover, the combination of the laser radiation and other non-optical processes, or the implementation of dual laser radiation systems, is a current requirement to improve both scientific reach and the quality of technical results. The prospective research in surface treatment by laser-assisted techniques appears to be directed towards short and ultra-short laser pulse texturing and the exploration of new laser wavelengths around the near-infrared and mid-infrared frontier.

Funding: This research received no external funding.

Acknowledgments: We would like to show our appreciation to all the authors, reviewers and editors for their contribution in this special issue of Coatings.

Conflicts of Interest: The authors declare no conflict of interest.

References

1. Liu, M.; Zhou, R.; Chen, Z.; Yan, H.; Cui, J.; Liu, W.; Pan, J.H.; Hong, M. Tunable hierarchical nanostructures on micro-conical arrays of laser textured TC4 substrate by hydrothermal treatment for enhanced anti-icing property. *Coatings* **2020**, *10*, 450. [CrossRef]

2. Torrisi, V.; Censabella, M.; Piccitto, G.; Compagnini, G.; Grimaldi, M.G.; Ruffino, F. Characteristics of Pd and Pt nanoparticles produced by nanosecond laser irradiations of thin films deposited on topographically-structured transparent conductive oxides. *Coatings* **2019**, *9*, 68. [CrossRef]

3. Fernández-Arias, M.; Zimbone, M.; Boutinguiza, M.; Del Val, J.; Riveiro, A.; Privitera, V.; Grimaldi, M.G.; Pou, J. Synthesis and deposition of Ag nanoparticles by combining laser ablation and electrophoretic deposition techniques. *Coatings* **2019**, *9*, 571. [CrossRef]

4. Baino, F.; Montealegre, M.A.; Minguella-Canela, J.; Vitale-Brovarone, C. Laser surface texturing of alumina/zirconia composite ceramics for potential use in hip joint prosthesis. *Coatings* **2019**, *9*, 369. [CrossRef]

5. Al-Sayyad, A.; Bardon, J.; Hirchenhahn, P.; Vaudémont, R.; Houssiau, L.; Plapper, P. Influence of aluminum laser ablation on interfacial thermal transfer and joint quality of laser welded aluminum–polyamide assemblies. *Coatings* **2019**, *9*, 768. [CrossRef]

6. Tayel, I.M. Thermoelastic response induced by volumetric absorption of uniform laser radiation in a half-space. *Coatings* **2020**, *10*, 228. [CrossRef]

7. Zhou, Y.; Zhao, Z.; Zhang, W.; Xiao, H.; Xu, X. Experiment study of rapid laser polishing of freeform steel surface by dual-beam. *Coatings* **2019**, *9*, 324. [CrossRef]

8. Penide, J.; del Val, J.; Riveiro, A.; Soto, R.; Comesaña, R.; Quintero, F.; Boutinguiza, M.; Lusquiños, F.; Pou, J. Laser surface blasting of granite stones using a laser scanning system. *Coatings* **2019**, *9*, 131. [CrossRef]

9. Wang, H.; Zhang, W.; Peng, Y.; Zhang, M.; Liu, S.; Liu, Y. Microstructures and wear resistance of FeCoCrNi-Mo high entropy alloy/diamond composite coatings by high speed laser cladding. *Coatings* **2020**, *10*, 300. [CrossRef]

10. Schopphoven, T.; Pirch, N.; Mann, S.; Poprawe, R.; Häfner, C.L.; Schleifenbaum, J.H. Statistical/numerical model of the powder-gas jet for extreme high-speed laser material deposition. *Coatings* **2020**, *10*, 416. [CrossRef]

11. Huang, K.; Li, W.; Pan, K.; Lin, X.; Wang, A. High temperature oxidation and thermal shock properties of $La_2Zr_2O_7$ thermal barrier coatings deposited on nickel-based superalloy by laser-cladding. *Coatings* **2020**, *10*, 370. [CrossRef]

© 2020 by the author. Licensee MDPI, Basel, Switzerland. This article is an open access article distributed under the terms and conditions of the Creative Commons Attribution (CC BY) license (http://creativecommons.org/licenses/by/4.0/).

Article

Tunable Hierarchical Nanostructures on Micro-Conical Arrays of Laser Textured TC4 Substrate by Hydrothermal Treatment for Enhanced Anti-Icing Property

Mengyao Liu [1,†], Rui Zhou [1,*,†], Zhekun Chen [1], Huangping Yan [1], Jingqin Cui [2], Wanshan Liu [1], Jia Hong Pan [3] and Minghui Hong [4,*]

[1] School of Aerospace Engineering, Xiamen University, Xiamen 361102, China; 19920171150953@stu.xmu.edu.cn (M.L.); 19920191151143@stu.xmu.edu.cn (Z.C.); hpyan@xmu.edu.cn (H.Y.); liuwanshan@xmu.edu.cn (W.L.)

[2] Pen-Tung Sah Institute of Micro-Nano Science and Technology, Xiamen University, Xiamen 361005, China; jqcui@xmu.edu.cn

[3] MOE Key Laboratory of Resources and Environmental Systems Optimization, College of Environmental Science and Engineering, North China Electric Power University, Beijing 102206, China; pan@ncepu.edu.cn

[4] Department of Electrical and Computer Engineering, National University of Singapore, Singapore 117576, Singapore

* Correspondence: rzhou2@xmu.edu.cn (R.Z.); elehmh@nus.edu.sg (M.H.)

† These authors contributed equally to this work.

Received: 11 March 2020; Accepted: 27 April 2020; Published: 6 May 2020

Abstract: In this work, an anti-icing structured surface was fabricated by combining laser ablation with hydrothermal treatment. A micro-patterned surface on a Ti alloy (TC4) substrate was easily fabricated by a highly effective nanosecond pulsed laser ablation. It was observed that titania (TiO$_2$) nanostructures were formed by hydrothermal treatment in aqueous alkali on the laser ablated TC4 substrate to obtain the micro/nano-hierarchical structures. The growth mechanism of the tunable nanoarrays was discussed by the adjustment of hydrothermal temperature. The as-prepared samples exhibited excellent superhydrophobicity with contact angles greater than 160°. It was found that optimized hydrothermal treatment on laser-processed TC4 substrates could further enhance surface anti-icing property. The results showed that the delay time (DT) had been extended by achieving over 90 min for the water droplets to freeze on the as-prepared structured surfaces, providing great potential in various anti-icing applications.

Keywords: laser ablation; hydrothermal treatment; micro/nano-hierarchical structures; wetting model; anti-icing

---

1. Introduction

Icing and ice accretion on aircraft structural material surfaces such as titanium alloys, typically of TC4 (Ti–6Al–4V) with good mechanical properties and high corrosion resistance, cause serious consequences such as freezing the wings, impairing sensing, and even ceasing operation [1,2]. There are a series of active solutions to solve aircraft icing such as thermal treatment and mechanical breaking. However, these methods incur plenty of energy waste and cause undesired damage to the main body. Hence, an alternative passive solution is in high demand so that ice will be prevented from accumulating on the anti-icing surfaces, which is advantageous in terms of energy and cost savings. With the fast development of nanotechnology and biomimetics, there have been many studies indicating that superhydrophobicity has positive effects on anti-icing property [3–5], although debates on the underlying mechanism are still unclear [6,7]. Additionally, no studies have clarified the comprehension of whether superhydrophobicity really suits icephobic applications by considering the durability issue of anti-icing property on a superhydrophobic surface due to the degradation of the coating materials [8,9]. Therefore, scientists have placed much attention on the surface structural properties of bulk materials and designing their superhydrophobic surface to reduce the possibility of surface icing.

Surface structure and surface energy are the two main factors affecting surface hydrophobicity. A large number of creatures in nature exhibit excellent hydrophobicity such as the micro-conical structure of the surface of pansy petals, which shows excellent hydrophobicity [10–12]. Furthermore, Herminghuas et al. found that hierarchical structures could make the surface wettability more significant [13]. In the past, researchers have successfully prepared superhydrophobic surfaces via fabricating surface structures through various methods [14] such as sandblasting [8], the anodization method [15], microwave irradiation [16], the sol-gel technique [17], chemical vapor deposition [18], and chemical etching [19]. However, it is quite hard to produce controllable biomimetic micro-array structures by these methods, generating harmful by-products for the environment. Instead, laser ablation is deemed as one of the most promising techniques to produce a periodic micro-array structure due to its high accuracy, environmental-friendly, controllability, and high efficiency in large-area manufacturing [20–25]. However, it was found that there are different sizes and disordered nanoparticles on the surfaces after laser ablation. These disordered nanostructures may cause defects, possibly weakening the de-wetting capability. Therefore, it is important to prepare tunable nanostructures on laser-induced microstructures. The hydrothermal reaction has been proposed as an excellent method to manufacture a large number of nanostructures. Moreover, it has been widely reported that nanostructures fabricated by hydrothermal reaction can be easily tuned by adjusting the hydrothermal temperature, leading to the formation of nanobelts, nanopetals, nanoneedles, and nanotubes [26–29]. However, the rational synthesis of a TiO_2 nanostructure on a laser ablated Ti alloy sheet by using the hydrothermal method has seldom been reported. This provided us with the inspiration to use nanosecond laser ablation to texture micron-scale structures on the surface and then hydrothermally grow nanoscale structures on the micron-scale structures, thereby constructing multiple hierarchical structures to enhance surface hydrophobicity.

In this work, laser processing was applied to fabricate the micro-conical array structure on a titanium alloy sheet, while hydrothermal treatment was adopted as a post-treatment process to grow and tune the nanoscale structure. After the surface texturing, the surface was modified by fluorosilane to obtain the superhydrophobic surfaces. The characteristics of the anti-icing properties are discussed for the as-prepared samples to reveal the underlying mechanisms. The results showed that the selected surfaces fabricated with the method above-mentioned could be equipped with enhanced anti-icing property compared to the sample only processed by laser or hydrothermal treatment. It is expected that this method will allow a bright future in designing better anti-icing surfaces.

2. Materials and Methods

2.1. Materials

TC4 Ti alloy sheets (Ti–6Al–4V, Shenzhen Hongwang Mold Co. Ltd., Shenzhen, China) with an area of 10 mm × 10 mm and a thickness of 1 mm were used as substrates. A Teflon-lined autoclave with 50 mL was used as the hydrothermal reaction container. Deionized water, ethanol, 1.0 mol/L NaOH solution, 0.8 mol/L HCl solution, and heptadecafluorodecyltrimethoxysilane (FAS-17, Qufu Jiaye Chemical New Material Co. Ltd., Qufu, China) were used for the following syntheses.

2.2. Sample Preparation

In this work, we used a Nd:YAG pulsed laser with a wavelength of 1064 nm and a pulse width of 100 ns. The spot size of the laser was 0.1 mm. The scanning speed, average laser power, and repetition frequency were set at 500 mm/s, 15 W, and 20 kHz, respectively, during the process. Prior to the laser texturing, the Ti alloy sheets were cleaned with ethanol and deionized water in an ultrasonic bath for 10 min sequentially and dried in air. Then, the Ti alloy sheets ablated by laser were transferred into a Teflon-lined autoclave containing 1.0 mol/L NaOH solution with 40 mL for hydrothermal treatment and the temperatures were set as 150, 170, 190, and 210 °C, respectively. After the hydrothermal reaction for 24 h, the obtained samples were immersed in a 0.8 mol/L HCl solution for 2 h by ion exchange reaction. Subsequently, the Ti alloy sheets were rinsed with deionized water and dried entirely. Then, the samples were subjected to annealing treatment at 300 °C for 3 h for the purpose of dehydration and annealing. Finally, to achieve the superhydrophobic surfaces, all samples were modified with 2 wt % solution of FAS-17/ethanol at ambient temperature for 6 h and well dried. A schematic diagram of the entire sample preparation process is shown in Figure 1. In addition, control samples were prepared. Control sample 1 is the substrate that was only processed by laser. Control sample 2 is the substrate without pre-treatment of laser ablation, which was only treated by hydrothermal reaction, ion exchange reaction, and annealing treatment with the same process. All control samples were covered with FAS-17.

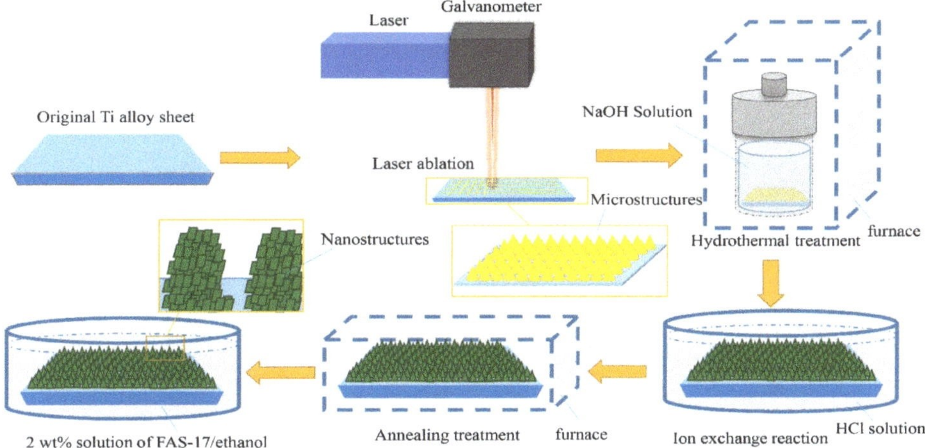

Figure 1. Schematic illustration of the fabrication process for the superhydrophobic Ti alloy.

2.3. Characterization

The as-prepared samples were observed with a Supra 55 scanning electron microscope (SEM, Carl Zeiss, Jena, Germany) for their surface morphologies and characterized with X-ray diffraction (XRD, Shimadzu, Kyoto, Japan) for possible phase and crystalline structure variations. The surface contact angle (CA) measurement was carried out in air at room temperature by a contact angle goniometer (KRUSS DSA-25, Hamburg, Germany) using the sessile drop method and 4 μL deionized water droplets. The CA for each sample was measured five times and the average value was adopted [22]. The surface water droplet adhesion test was carried out by using 6 μL of deionized water droplets.

2.4. Anti-Icing Property

The in-house anti-icing testing platform was established and included a temperature control platform, camera, and a computer. The surface temperature of the experimental platform could be tuned and stabilized at −5 °C under ambient conditions, fully supporting the characterization of anti-icing property. First, the as-prepared sample was fixed with heat-conducting silicone grease on the experimental plate horizontally. Second, deionized water droplets with the volume of 6 μL were dropped onto the surfaces. Then, the temperature of the experimental platform was decreased from room temperature to −5 °C with a decreasing rate of 1 °C/s by a temperature controller. The temperature of the plate could be in-situ monitored by a digital temperature measuring instrument and fed back to the temperature controller, ensuring that the temperature stabilizes at −5 °C throughout the experiment. The time recording starts when the plate temperature reaches −5 °C. Meanwhile, the icing process is monitored and recorded by the camera. The delay time, DT (i.e., the time taken for the deionized water droplets to freeze), is recorded by observing the non-transparency of the droplets on the surface at −5 °C. In this work, DT is the measurement of the sample's anti-icing property [30]. The anti-icing characterization for all samples were measured five times. Since the desired micro-conical arrays do not exist for control sample 2 without pre-treatment of laser ablation, the control experiment of anti-icing characterization was only carried out on control sample 1, which was only processed by laser.

3. Results and Discussion

The large-area SEM images of the surfaces of the micro/nano-hierarchical structures with micro-conical arrays synthesized by the combination of laser process and hydrothermal reaction at 150 °C are shown in Figure 2a–c. It can be seen that a large number of small micro-conical array structures with the same sizes were evenly located on the as-prepared sample surface (Figure 2a), and those cones were similar to the micro-structure pattern of the surface of the pansy petal [12]. A closer examination revealed that micro-cones were vertically located in the as-prepared substrate with diameters of about 200 μm at the bottom. The micro-conical array structure was produced by laser ablation. When the nanosecond pulsed laser hits the surface of the target material, a large number of photons are absorbed by the electrons of the TC4 substrate and transferred to the crystal lattice in a short time (~picosecond level), causing thermal diffusion and raising the temperature of the target material, leading to heating, melting, and evaporation [31]. The heat-affected zone can change the surface morphology by recasting and formation of burrs. In laser ablation, compressive forces can be created by the vapor and plasma plume at the focus point, leading to the expulsion of a molten pool of materials [32]. A portion of the liquefied material is resolidified after pressed out of the ablation zone, and accumulates and covers the surrounding non-processed areas, eventually forming a conical structure. Figure 2b is the top-view SEM image of the micro-conical structure. With further investigation, it can be observed that the surface was covered by many nanoscale flocculent structures, as shown in Figure 2c. These flocculent structures are synthesized by the hydrothermal reaction and are composed of layered unit cells. The crystalline structure of the flocs can be described with a representative TiO_6 octahedral. During the alkali-hydrothermal process, the laser processed titanium alloy reacts with the NaOH solution at 150 °C in the autoclave, where some of the Ti–O–Ti bonds are broken and form the six-coordinated monomer $[Ti(OH)_6]^{2-}$, which can form polynuclear complexes by olation or oxolation, depending on the concentration of the solution at different temperatures. Under the hydrothermal process, the solution is in a saturated state; hence, the $[Ti(OH)_6]^{2-}$ is unstable and tends to combine via oxolation or olation, forming original nuclei [33,34]. As these nuclei grow and exceed the critical nuclei size, they become stable. Small and thin flocs can be formed when the growth continues [33]. Eventually, it can be observed that the micro/nano-hierarchical structures were successfully fabricated on the Ti alloy sheet surface. The surface of the Ti alloy is initially hydrophilic since the Ti alloy is a kind of material with high surface energy. According to the surface structural influence on the surface wettability, the surface structure can make a hydrophilic surface more hydrophilic, and a hydrophobic surface more hydrophobic. Thus, the surface of the Ti alloy with micro/nano-hierarchical structures becomes a superhydrophilic surface. Micro-structure and low surface energy are both prerequisites to achieve surface superhydrophobicity [35]. Therefore, in order to obtain superhydrophobic surfaces, samples were immersed inside the FAS-17 solution for low surface energy modification. After the treatment, the water droplet exhibited a spherical shape on the sample surface with a contact angle of approximately 160° ± 2°, as illustrated in Figure 2e. Furthermore, the as-prepared surface was very slippery and it was easy for the water droplets to slide off with an inclination angle of less than 1°, which is shown in Figure 2f. This indicates that the solid–liquid wetting model between the water droplet and the sample surface is in Cassie–Baxter state, as shown as Figure 2d.

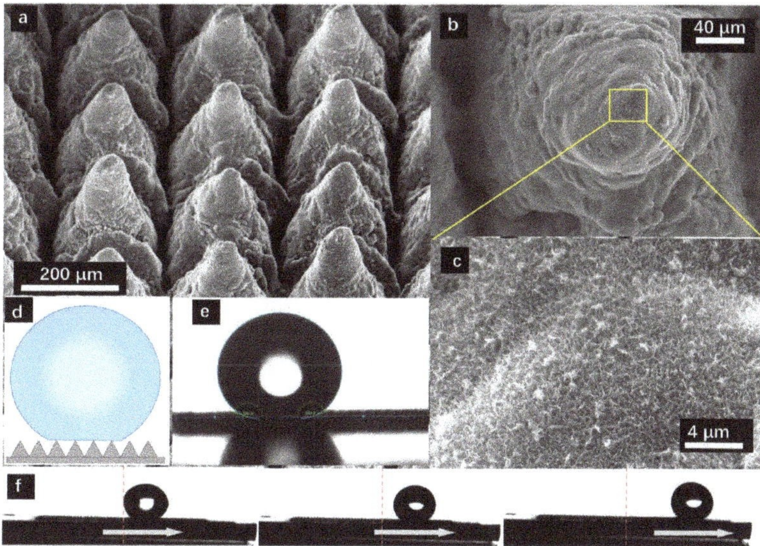

Figure 2. Scanning electron microscopy (SEM) images of the as-prepared surface with different magnifications: (**a**) 100x; (**b**) 500x; (**c**) 5000x; (**d**) Cassie–Baxter state; (**e**) Optical micrograph of water droplets; (**f**) Water droplets slide off the surface. Hydrothermal temperature is 150 °C.

In this work, the micro-conical structures of TiO_2 were first created on the TC4 substrates by laser ablation. The secondary nanostructure was then fabricated by imposing a hydrothermal treatment at different temperatures. According to the XRD pattern shown in Figure 3a, after the hydrothermal reaction, the samples exhibited the strong peaks of the (110), ($21\bar{1}$), (121), and (431) planes, indicating that Na_2TiO_3 nanostructures had been successfully synthesized during the alkali-hydrothermal process, corresponding to Equation (1).

$$TiO_2 + 2NaOH = Na_2TiO_3 + H_2O \tag{1}$$

By increasing the temperature from 150 to 210 °C, the intensities of the (110) and (121) peaks increased, indicating the gradual enhancement in the crystallinity of the as-prepared Na_2TiO_3 nanostructures. After soaking and washing in HCl solution, the Na^+ ions were replaced by H^+ ions to produce H_2TiO_3 nanostructures, corresponding to Equation (2).

$$Na_2TiO_3 + 2HCl = H_2TiO_3 + 2NaCl \text{ (ion exchange reaction)} \tag{2}$$

Finally, a calcination process was used for the dehydration of H_2TiO_3 to obtain crystalline TiO_2 nanostructures, corresponding to Equation (3).

$$H_2TiO_3 = TiO_2 + H_2O(g) \text{ (annealing treatment)} \tag{3}$$

The XRD patterns of the calcined samples are shown in Figure 3b. The diffraction peaks of the (101), (110), and (112) planes can be well indexed as TiO_2. By increasing the temperature from 150 to 210 °C, the intensities of the diffraction peaks increased, indicating that the crystallinity of the as-prepared TiO_2 nanostructures gradually enhanced.

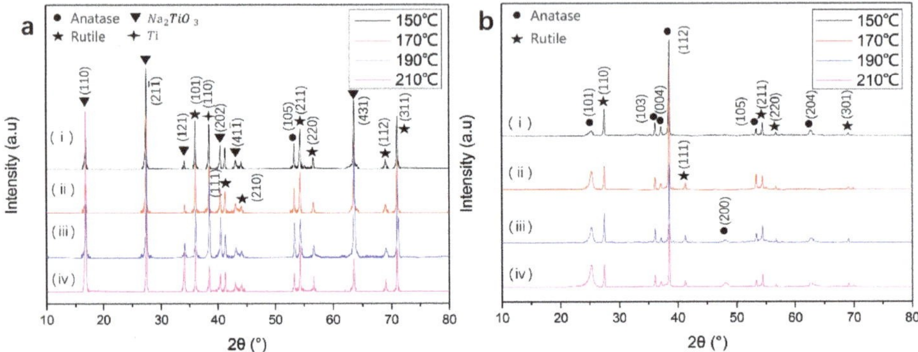

Figure 3. X-ray diffraction (XRD) patterns with different hydrothermal temperatures of micro/nano-hierarchical structures: (i) Surf 1: 150 °C; (ii) Surf 2: 170 °C; (iii) Surf 3: 190 °C; (iv) Surf 4: 210 °C. (**a**) After hydrothermal reaction; (**b**) After annealing treatment at 300 °C for 3 h.

As shown in Figure 4, it can be observed that the nanostructures on the micro-conical array produced by the hydrothermal treatment could be tuned by adjusting the hydrothermal temperature. It could also be found that these nanostructures were smaller than the particle, which were covered on control sample 1. Furthermore, as seen in control sample 2, it only produced large cluster structures when compared to the refined nanostructures by the hydrothermal treatment of the laser textured samples. The laser-ablated surface of the sample can form small needles or filamentous structures after hydrothermal treatment. This may be due to the easier growth of nanostructured crystals by hydrothermal processes on TiO_2 obtained after laser ablation. At the beginning of the hydrothermal reaction, TiO_2 reacted with the NaOH aqueous solution and many TiO_6 octahedral ions were created [36]. Then, the TiO_6 octahedral ions electrostatically combined with Na^+ ions and OH^- ions to form Na_2TiO_3 crystal nuclei [33]. During the hydrothermal reaction, the nucleus are randomly distributed on the surface of the laser-treated titanium alloy substrate and grow to form nanostructures [28,36]. Therefore, only the combination of laser ablation and the hydrothermal method can produce ideal micro/nano-hierarchical structures. The images of Figure 4a–d correspond to hydrothermal temperatures of 150, 170, 190, and 210 °C, respectively. As shown in Figure 4a, when the hydrothermal temperature was 150 °C, Surf 1 was covered by flocculent nanostructures with poor crystallinity. Many crystal dislocations/defects could be produced, existing at the crystal boundary during the hydrothermal growth process. When the hydrothermal temperature increased, the number of crystal nuclei generated by the hydrothermal reaction increased, and the rate of crystal growth also increased. The crystalline growth in the cross-sectional direction was inhibited due to the competitive growth from a large number of crystals simultaneously, leading to the inner stress in the radial direction [26,36]. While increasing the temperature up to 170 °C, the inner stress could suppress the crystal to form nanofloc. The nanostructures began to grow vertically on the substrate through a dissolution–recrystallization process, where needle-like nanostructures appeared on the surface of the sample (as shown in Figure 4b, Surf 2). When the hydrothermal temperature reached 190 °C, the length of the nanoneedles could be increased by elevating the reaction temperature [26], contributing to the formation of filamentous structures on the surface of the sample (Figure 4c, Surf 3). By further raising the hydrothermal temperature to 210 °C, it can be observed that the wire structured crystal could be achieved (Figure 4d, Surf 4). In this study, it was found that at a low reaction temperature, the nanostructures were mainly short and flat nanofloc. When the temperature increased, the nanosheets not only elongated, but are also shrank in the cross direction. At a sufficiently high temperature, the nanostructures became filamentous.

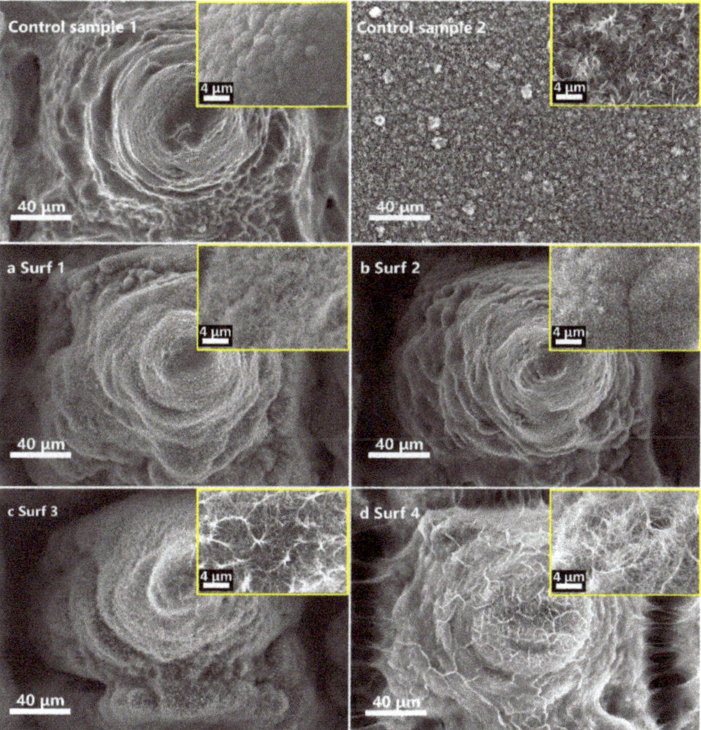

Figure 4. Control sample 1 is the sample only treated by laser processing. Control sample 2 is the sample only treated by hydrothermal treatment, where the hydrothermal temperature was 190 °C. (**a**–**d**) SEM images of the hydrothermal reaction Ti alloy surfaces after laser ablation with a magnification of 500x. The hydrothermal temperature is (**a**) Surf 1: 150 °C; (**b**) Surf 2: 170 °C; (**c**) Surf 3: 190 °C; (**d**) Surf 4: 210 °C. Inset images is at the magnification of 5000x.

Table 1 summarizes the water contact angles (WCA) of all samples with micro/nano-hierarchical surfaces, which were modified by FAS-17. The WCAs of all surfaces exceeded 160° in their natural state at room temperature and ambient pressure, indicating the superhydrophobic surfaces. It is well-known that there are two classic models widely used to explain the hydrophobicity of rough surfaces: the Cassie–Baxter state and the Wenzel state [37]. When water droplets cannot penetrate into the structure, the solid–liquid contact state can be assumed as the Cassie–Baxter state. In this condition, the air trapped in the structured gaps can block the contact between the liquid and the surface, resulting in the water droplets easily rolling off from the surface without sticking. It also promotes a reduction in the contact area between the droplet and the sample surface. When the solid–liquid wetting model is in the Wenzel state, the surface still exhibits a large contact angle, however, water droplets are significantly infiltrated into the surface structure, so the water droplets adhere on the sample and it is hard to roll off them from the surface. Therefore, the solid–liquid contact state could only be determined by the combination of static contact angle measurement and the adhesion test of the surface to water droplets in the case of superhydrophobicity. Figure 5 shows the adhesion test of water droplets on the sample surface. The water droplet bumped on all surfaces and then left off. It can be found that the water droplets did not adhere onto the surfaces, which means that these micro/nano-hierarchical structured patterns can perform a Cassie–Baxter state when they make contact with the water droplets, leading to the blocking of water droplets to penetrate into the pattern and be suspended on the as-prepared surfaces [38–40].

Table 1. Water contact angle of all micro/nano-hierarchical surfaces.

Hydrothermal Temperature/°C	150	170	190	210
Samples	Surf 1	Surf 2	Surf 3	Surf 4
WCA/°	161.2 ± 1.3	161.9 ± 1.3	161.7 ± 1.1	161.9 ± 0.4

Figure 5. Surface water droplet adhesion test corresponding to different hydrothermal temperatures of micro/nano hierarchical structures: (i) Surf 1: 150 °C; (ii) Surf 2: 170 °C; (iii) Surf 3: 190 °C; (iv) Surf 4: 210 °C.

4. Anti-Icing Property

The anti-icing property of the as-prepared surfaces was investigated by the in-house platform. Figure 6a shows the photographs of the icing process of deionized water droplets on the as-prepared surfaces. Surf 0 is the control sample 1 where the surface was only processed by laser. Initially, the reference droplets on all surfaces were transparent (Frame 1). Then, it was observed that the droplet on Surf 4 became non-transparent after ~17 min (frame 2), which indicated that the droplet was frozen. After ~34 min (frame 3), the droplet on Surf 0 became non-transparent. In contrast, the droplets on other textured surfaces were still transparent. Then, after ~95 min (Frames 5 and 6), all the water droplets on the rest of surfaces were frozen. This implies that these surfaces (e.g., Surf 1, Surf 2, and Surf 3) had a relatively long time to resist the water freezing. Obviously, the micro/nano-hierarchical structure synthesized by the collaborative effect of laser processing and hydrothermal treatment at a suitable temperature can effectively enhance the surface anti-icing property, in comparison with the sample processed only by laser. Water droplets need to lose heat to be frozen. When the temperature of a water droplet decreases below the freezing point, the molecules of water tend to bond in an orderly arrangement to form an ice nucleus and be frozen. During the anti-icing characterization, the water droplet loses heat to the samples' surface. According to the analysis of structural characteristics, the nanostructures synthesized by hydrothermal treatment take up a part of the surface layer and a lot of air is trapped in the nanostructure to form air bags [38,39]. Thus, the water droplet is suspended up onto the micro-cone with nanostructures to decrease the wetting contact with the solid surface. These air bags rarely occur on surfaces with only micro-structures. Therefore, on the surface of the micro/nano-hierarchical structure, the water droplets isolated by the air bags find it more difficult to transfer the heat to the surface of the sample, resulting in the prolonged freezing time of water droplets.

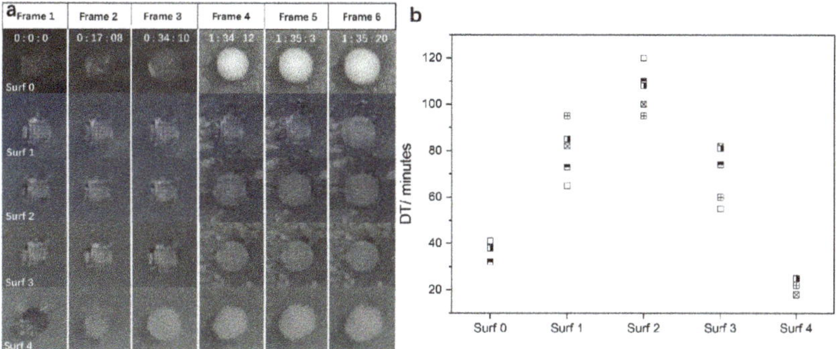

Figure 6. In-situ observation of ice formation on the as-prepared surfaces at −5 °C with delay time (DT): (**a**) Images of the water droplet icing on the as-prepared surfaces along with time prolonged; (**b**) The delay time (DT) measured five times for all samples.

After multiple measurements of anti-icing characterization, it can be observed that, when the hydrothermal temperature is 150, 170 and 190 °C, the anti-icing property can be really enhanced via tunable hierarchical nanostructures on micro-conical arrays of laser ablated substrates by hydrothermal treatment. However, it is interestingly that the DT of Surf 4 was much shorter than that of other surfaces, which means that the anti-icing property of Surf 4 becomes worse, as shown in Figure 6b. As known, the process of icing is mainly affected by the heat transfer between water and the medium. In terms of thermodynamics, the icing of water droplets needs to transfer heat to air and the substrate material under the temperature gradient. In this work, the water droplets were suspended over the as-prepared superhydrophobic surfaces and the solid–liquid–air three-phase interfaces existed, providing the droplet with the routes to gain or lose heat. For example, while decreasing the temperature of the anti-icing characterization platform, the water droplet loses heat to the cold as-prepared surfaces through contact heat conduction and thermal radiation between the droplet and the surface. On the other hand, the water droplets gain heat from air in the form of thermal radiation [41]. From the aspect of energy equilibrium, it can be quantitatively described as Equation (4) on the cold as-prepared surfaces:

$$Q_d = Q_o + Q_g + Q_l,\tag{4}$$

where Q_d is the heat energy of droplet, which corresponds to its enthalpy; Q_o is the original heat energy of droplet; Q_g is the heat energy gains from air; and Q_l is the heat energy loss by thermal radiation between the droplet and the surface. The droplet is hypothesized to be stable in a spherical shape and the heat transfer from any point around the water droplet is homogeneous. The description of the heat transfer between the droplet and ambient air is as per Equation (5):

$$Q_g = \alpha S_d (T_a - T_d),\tag{5}$$

where α is the radiant heat-transfer coefficient; S_d is the surface area of the droplet; T_a is the temperature of ambient air; and T_d is the temperature of the droplet. Fourier's law is also introduced to describe the heat transfer (Q_l) between the droplet and surface as Equation (6):

$$Q_l = \int_{T_1}^{T_2} K S_c dT,\tag{6}$$

where T_1 is the temperature of the droplet; T_2 is the temperature of the sample surface, which is controlled by the temperature control platform; K is the heat-transfer coefficient; and S_c is the solid–liquid contact area. Combining Equations (4)–(6), the heat transfer equation is derived as Equation (7):

$$Q_d = Q_o + \alpha S_d (T_a - T_d) + \int_{T_1}^{T_2} K S_c dT, \tag{7}$$

Assuming that the water droplet on the as-prepared surfaces lose the same amount of heat to air, the contact heat conduction from water droplets to the sample surfaces is the key factor affecting the DT of droplets. Thus, the solid–liquid contact area is the key factor affecting the contact heat conduction, influencing the DT of droplets. The effect of different morphologies on the solid–liquid contact area is further investigated by observing the contact states of various surfaces with water droplets after cooling. After placing all the micro/nano-hierarchical structure samples at −5 °C for 3 min, it was observed that the water droplet on Surf 4 was firmly adhered to the surface, but the water droplets on other surfaces easily rolled off, as shown in Figure 7a, indicating that the superhydrophobic state of Surf 4 may change from a Cassie–Baxter state to a Wenzel state in 3 min, while Surf 1–3 still maintained a Cassie–Baxter state. This may be due to the condensation of the humidity on the surface, which occurred at low temperature. Surf 4 seems to have a much larger surface area than the other samples (because it was covered with numerous nanowires, as shown in Figure 4), which increased the number of van der Waals interactions between the vapor phase molecules inside the confined spaces and is more exposed to the water capillary condensation phenomenon that leads to the loss of the Cassie–Baxter state, allowing water droplets to penetrate into the surface and forming "sticky" drops. When Surf 4 was subsequently tilted, the droplet remained adhered without rolling-off behavior [42]. This greatly increased the solid–liquid contact area S_c (as shown in Figure 7b), enhancing the heat dissipation efficiency of the liquid and causing the water droplets on the surface to freeze more easily. Accordingly, it was observed that $t_{surf\,4} < t_{surf\,0} < t_{surf\,3} < t_{surf\,1} < t_{surf\,2}$, where t is the duration to sustain the Cassie–Baxter state before state transition at low temperature, resulting in $DT_{surf\,4} < DT_{surf\,0} < DT_{surf\,3} < DT_{surf\,1} < DT_{surf\,2}$, as shown in Figure 6b. Consequently, the structural differences of the superhydrophobic surface can significantly impact delayed freezing time. The capability of the as-prepared surfaces to maintain the superhydrophobic Cassie–Baxter state determines the delayed freezing time. Although the micro/nano-hierarchical structures can effectively delay the freezing time, water droplets are more likely to penetrate into the structure and accelerate freezing while the nanowire structure dominates the surface morphology.

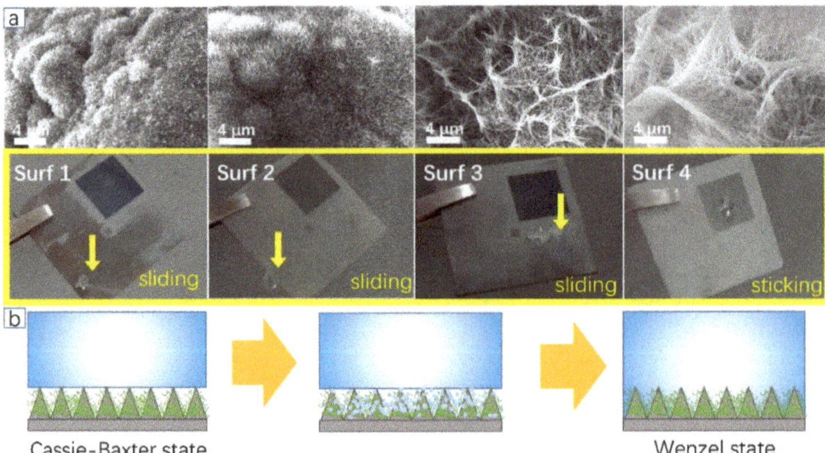

Figure 7. (**a**) Water droplet adhesion test after 3 min at a temperature of −5 °C; (**b**) The abridged general view of transformation from the Cassie–Baxter state to the Wenzel state by decreasing the ambient temperature.

5. Conclusions

Superhydrophobic surfaces with hierarchical nanostructures on micro-conical arrays have been successfully fabricated on Ti alloy substrates through a combination of laser ablation and hydrothermal treatment. The nanostructures, transforming from flocs to filamentous, can be tuned by adjusting the hydrothermal temperature. Moreover, the constructed superhydrophobic surfaces with a micro/nano-hierarchical structure can efficiently delay the freezing of water droplets, which is mainly attributed to the fact that the structure hindered the contact of the water droplets with the solids, reducing the heat loss efficiency of the water droplets and enabling the resistance of water droplets for a longer time without freezing. It was also found that denser nanowire structures are likely to trap the water droplets and convert the solid–liquid wetting model from the Cassie-Baxter to Wenzel states by decreasing the ambient temperature, which shortens the delay time of the water droplet freezing. It is expected that micro/nano-hierarchical structures with nanoneedles on micro-conical arrays promise an achievable future for developing a surface anti-icing strategy.

Author Contributions: Conceptualization, M.L. and R.Z.; Data Curation, M.L. and Z.C.; Formal Analysis, R.Z., H.Y., J.C., and J.H.P.; Funding Acquisition, R.Z.; Investigation, M.L. and Z.C.; Methodology, M.L. and R.Z.; Project administration, R.Z.; Resources, R.Z.; Software, M.L.; Supervision, R.Z. and M.H.; Validation, Z.C. and W.L.; Visualization, M.L. and R.Z.; Writing–Original Draft, M.L. and R.Z.; Writing–Review and Editing, M.L., R.Z., and M.H. All authors have read and agreed to the published version of the manuscript.

Funding: This research was funded by the Ministry of Education of China, Grant NO. CXZJHZ201738.

Acknowledgments: We acknowledge the support of the Collaborative Innovation Center of High-End Equipment Manufacturing in Fujian.

Conflicts of Interest: The authors declare no conflicts of interest. The funders had no role in the design of the study; in the collection, analyses, or interpretation of data; in the writing of the manuscript, or in the decision to publish the results.

References

1. Cao, Y.H.; Tan, W.Y.; Wu, Z.L. Aircraft icing: An ongoing threat to aviation safety. *Aerosp. Sci. Technol.* **2018**, *75*, 353–385. [CrossRef]
2. Bin, Y.; Yanpei, Z. Icing Certification of Civil Aircraft Engines. *Procedia Eng.* **2011**, *17*, 603–615. [CrossRef]
3. Zhang, H.Y.; Yang, Y.L.; Pan, J.F.; Long, H.; Huang, L.S.; Zhang, X.K. Compare study between icephobicity and superhydrophobicity. *Phys. B Condens. Matter* **2019**, *556*, 118–130. [CrossRef]
4. Zhang, Z.; Liu, X.Y. Control of ice nucleation: Freezing and antifreeze strategies. *Che. Soc. Rev.* **2018**, *47*, 7116–7139. [CrossRef]
5. Farhadi, S.; Farzaneh, M.; Kulinich, S.A. Anti-icing performance of superhydrophobic surfaces. *Appl. Surf. Sci.* **2011**, *257*, 6264–6269. [CrossRef]
6. Nosonovsky, M.; Hejazi, V. Why Superhydrophobic Surfaces Are Not Always Icephobic. *ACS Nano* **2012**, *6*, 8488–8491. [CrossRef]
7. Kreder, M.J.; Alvarenga, J.; Kim, P.; Aizenberg, J. Design of anti-icing surfaces: Smooth, textured or slippery? *Nat. Rev. Mater.* **2016**, *1*. [CrossRef]
8. Balordi, M.; Cammi, A.; Santucci de Magistris, G.; Chemelli, C. Role of micrometric roughness on anti-ice properties and durability of hierarchical super-hydrophobic aluminum surfaces. *Surf. Coat. Technol.* **2019**, *374*, 549–556. [CrossRef]
9. Wang, Y.; Xue, J.; Wang, Q.; Chen, Q.; Ding, J. Verification of Icephobic/Anti-icing Properties of a Superhydrophobic Surface. *ACS Appl. Mater. Interfaces* **2013**, *5*, 3370–3381. [CrossRef]
10. Han, Z.; Mu, Z.; Yin, W.; Li, W.; Niu, S.; Zhang, J.; Ren, L. Biomimetic multifunctional surfaces inspired from animals. *Adv. Colloid Interface Sci.* **2016**, *234*, 27–50. [CrossRef]
11. Liu, M.-N.; Wang, L.; Yu, Y.-H.; Li, A.-W. Biomimetic construction of hierarchical structures via laser processing. *Opt. Mater. Express* **2017**, *7*, 2208. [CrossRef]
12. Koch, K.; Bhushan, B.; Barthlott, W. Multifunctional surface structures of plants: An inspiration for biomimetics. *Prog. Mater. Sci.* **2009**, *54*, 137–178. [CrossRef]
13. Herminghaus, S. Roughness-induced non-wetting. *Europhys. Lett.* **2000**, *52*, 165–170. [CrossRef]

14. Gurera, D.; Bhushan, B. Fabrication of bioinspired, self-cleaning, anti-icing, superliquiphilic/phobic titanium using different pathways. *Philos. Trans. R. Soc. A* **2018**, *377*, 20180273. [CrossRef]

15. Li, S.-Y.; Li, Y.; Wang, J.; Nan, Y.-G.; Ma, B.-H.; Liu, Z.-L.; Gu, J.-X. Fabrication of pinecone-like structure superhydrophobic surface on titanium substrate and its self-cleaning property. *Chem. Eng. J.* **2016**, *290*, 82–90. [CrossRef]

16. Lee, E.; Lee, K.H. Facile fabrication of superhydrophobic surfaces with hierarchical structures. *Sci. Rep. UK* **2018**, *8*, 4101. [CrossRef]

17. Tang, Y.; Zhang, Q.; Zhan, X.; Chen, F. Superhydrophobic and anti-icing properties at overcooled temperature of a fluorinated hybrid surface prepared via a sol-gel process. *Soft Matter* **2015**, *11*, 4540–4550. [CrossRef]

18. Zhou, B.; Tian, J.; Wang, C.; Gao, Y.; Wen, W. A facile and cost-effective approach to engineer surface roughness for preparation of large-scale superhydrophobic substrate with high adhesive force. *Appl. Surf. Sci.* **2016**, *389*, 679–687. [CrossRef]

19. Ganne, A.; Lebed, V.O.; Gavrilov, A.I. Combined wet chemical etching and anodic oxidation for obtaining the superhydrophobic meshes with anti-icing performance. *Colloids Surf. A* **2016**, *499*, 150–155. [CrossRef]

20. Yan, H.; Abdul Rashid, M.R.B.; Khew, S.Y.; Li, F.; Hong, M. Wettability transition of laser textured brass surfaces inside different mediums. *Appl. Surf. Sci.* **2018**, *427*, 369–375. [CrossRef]

21. Pou, P.; del Val, J.; Riveiro, A.; Comesaña, R.; Arias-González, F.; Lusquiños, F.; Bountinguiza, M.; Quintero, F.; Pou, J. Laser texturing of stainless steel under different processing atmospheres: From superhydrophilic to superhydrophobic surfaces. *Appl. Surf. Sci.* **2019**, *475*, 896–905. [CrossRef]

22. Ta, V.D.; Dunn, A.; Wasley, T.J.; Li, J.; Kay, R.W.; Stringer, J.; Smith, P.J.; Esenturk, E.; Connaughton, C.; Shephard, J.D. Laser textured superhydrophobic surfaces and their applications for homogeneous spot deposition. *Appl. Surf. Sci.* **2016**, *365*, 153–159. [CrossRef]

23. Yan, H.; Xiao, X.; Chen, Z.; Chen, Y.; Zhou, R.; Wang, Z.; Hong, M. Realization of adhesion enhancement of CuO nanowires growth on copper substrate by laser texturing. *Opt. Laser Technol.* **2019**, *119*, 105612. [CrossRef]

24. Khew, S.Y.; Tan, C.F.; Yan, H.; Lin, S.; Thian, E.S.; Zhou, R.; Hong, M. Nanosecond laser ablation for enhanced adhesion of CuO nanowires on copper substrate and its application for oil-water separation. *Appl. Surf. Sci.* **2019**, *465*, 995–1002. [CrossRef]

25. Zhou, R.; Lu, X.Z.; Lin, S.D.; Huang, T.T. Laser Texturing of NiTi Alloy with Enhanced Bioactivity for Stem Cell Growth and Alignment. *J. Laser Micro/Nanoeng.* **2017**, *12*, 22–27. [CrossRef]

26. Wang, W.; Lin, H.; Li, J.; Wang, N. Formation of Titania Nanoarrays by Hydrothermal Reaction and Their Application in Photovoltaic Cells. *J. Am. Ceram. Soc.* **2008**, *91*, 628–631. [CrossRef]

27. Guo, H.; Ding, R.; Li, N.; Hong, K.; Liu, L.; Zhang, H. Defects controllable ZnO nanowire arrays by a hydrothermal growth method for dye-sensitized solar cells. *Phys. E* **2019**, *105*, 156–161. [CrossRef]

28. Li, Y.; Guo, M.; Zhang, M.; Wang, X. Hydrothermal synthesis and characterization of TiO_2 nanorod arrays on glass substrates. *Mater. Res. Bull.* **2009**, *44*, 1232–1237. [CrossRef]

29. Zuo, Z.; Liao, R.; Zhao, X.; Song, X.; Qiao, Z.; Guo, C.; Zhuang, A.; Yuan, Y. Anti-frosting performance of superhydrophobic surface with ZnO nanorods. *Appl. Therm. Eng.* **2017**, *110*, 39–48. [CrossRef]

30. Guo, P.; Zheng, Y.; Wen, M.; Song, C.; Lin, Y.; Jiang, L. Icephobic/anti-icing properties of micro/nanostructured surfaces. *Adv. Mater.* **2012**, *24*, 2642–2648. [CrossRef]

31. Wang, Z.; Zhou, R.; Wen, F.; Zhang, R.; Ren, L.; Teoh, S.H.; Hong, M. Reliable laser fabrication: The quest for responsive biomaterials surface. *J. Mater. Chem. B* **2018**, *6*, 3612–3631. [CrossRef] [PubMed]

32. Zhou, R.; Lin, S.; Ding, Y.; Yang, H.; Ong Yong Keng, K.; Hong, M. Enhancement of laser ablation via interacting spatial double-pulse effect. *Opto-Electron. Adv.* **2018**, *1*, 18001401–18001406. [CrossRef]

33. Wang, Y.M.; Du, G.J.; Liu, H.; Liu, D.; Qin, S.B.; Wang, N.; Hu, C.G.; Tao, X.T.; Jiao, J.; Wang, J.Y.; et al. Nanostructured sheets of Ti-O nanobelts for gas sensing and antibacterial applications. *Adv. Funct. Mater.* **2008**, *18*, 1131–1137. [CrossRef]

34. Yanqing, Z.; Erwei, S.; Zhizhan, C.; Wenjun, L.; Xingfang, H. Influence of solution concentration on the hydrothermal preparation of titania crystallites. *J. Mater. Chem.* **2001**, *11*, 1547–1551. [CrossRef]

35. Tang, M.; Shim, V.; Pan, Z.Y.; Choo, Y.S.; Hong, M.H. Laser Ablation of Metal Substrates for Super-hydrophobic Effect. *J. Laser/Micro Nanoeng.* **2011**, *6*, 6–9. [CrossRef]

36. Liu, W.; Lu, H.; Zhang, M.; Guo, M. Controllable preparation of TiO_2 nanowire arrays on titanium mesh for flexible dye-sensitized solar cells. *Appl. Surf. Sci.* **2015**, *347*, 214–223. [CrossRef]

37. Erbil, H.Y.; Cansoy, C.E. Range of applicability of the Wenzel and Cassie-Baxter equations for superhydrophobic surfaces. *Langmuir* **2009**, *25*, 14135–14145. [CrossRef]
38. Yang, Q.; Luo, Z.; Jiang, F.; Luo, Y.; Tan, S.; Lu, Z.; Zhang, Z.; Liu, W. Air Cushion Convection Inhibiting Icing of Self-Cleaning Surfaces. *ACS Appl. Mater. Interfaces* **2016**, *8*, 29169–29178. [CrossRef]
39. Gao, Y.; Sun, Y.; Guo, D. Facile fabrication of superhydrophobic surfaces with low roughness on Ti–6Al–4V substrates via anodization. *Appl. Surf. Sci.* **2014**, *314*, 754–759. [CrossRef]
40. Li, B.-J.; Li, H.; Huang, L.-J.; Ren, N.-F.; Kong, X. Femtosecond pulsed laser textured titanium surfaces with stable superhydrophilicity and superhydrophobicity. *Appl. Surf. Sci.* **2016**, *389*, 585–593. [CrossRef]
41. Zhan, X.; Yan, Y.; Zhang, Q.; Chen, F. A novel superhydrophobic hybrid nanocomposite material prepared by surface-initiated AGET ATRP and its anti-icing properties. *J. Mater. Chem. A* **2014**, *2*, 9390–9399. [CrossRef]
42. Cheng, Y.-T.; Rodak, D.E. Is the lotus leaf superhydrophobic? *Appl. Phys. Lett.* **2005**, *86*, 144101. [CrossRef]

© 2020 by the authors. Licensee MDPI, Basel, Switzerland. This article is an open access article distributed under the terms and conditions of the Creative Commons Attribution (CC BY) license (http://creativecommons.org/licenses/by/4.0/).

Article

Characteristics of Pd and Pt Nanoparticles Produced by Nanosecond Laser Irradiations of Thin Films Deposited on Topographically-Structured Transparent Conductive Oxides

Vanna Torrisi [1,*], **Maria Censabella** [2], **Giovanni Piccitto** [2], **Giuseppe Compagnini** [3], **Maria Grazia Grimaldi** [2] and **Francesco Ruffino** [2,*]

1   BRIT (Bio-Nanotech Research Innovation Tower), Università di Catania, via S. Sofia 89, 95123 Catania, Italy
2   Dipartimento di Fisica e Astronomia (Ettore Majorana), Università di Catania and MATIS CNR-IMM, via S. Sofia 64, 95123 Catania, Italy; maria.censabella@ct.infn.it (M.C.); giovanni.piccitto@ct.infn.it (G.P.); mariagrazia.grimaldi@ct.infn.it (M.G.G.)
3   Dipartimento di Scienze Chimiche, Università di Catania, Viale A. Doria 6, 95125 Catania, Italy; gcompagnini@dipchi.unict.it
*   Correspondence: v.torrisi@unict.it (V.T.); francesco.ruffino@ct.infn.it (F.R.)

Received: 10 November 2018; Accepted: 22 January 2019; Published: 24 January 2019

Abstract: Pd and Pt nanoparticles on Fluorine-doped tin oxide (FTO) are produced. This outcome is reached by processing nanoscale-thick Pd and Pt films deposited on the FTO surface by nanosecond laser pulse. Such laser processes are demonstrated to initiate a dewetting phenomenon in the deposited metal films and lead to the formation of the nanoparticles. In particular, the effect of the film's thickness on the mean size of the nanoparticles, when fixed the laser fluence, is studied. Our results indicate that the substrate topography influences the dewetting process of the metal films and, as a consequence, impacts on the nanoparticle characteristics. The results concerning the Pd and Pt nanoparticles' sizes versus starting films thickness and substrate topography are discussed. In particular, the presented discussion is based on the elucidation of the effect of the substrate topography effect on the dewetting process through the excess of chemical potential. Finally, Raman analysis on the fabricated samples are presented. They show, in particular for the case of the Pd nanoparticles on FTO, a pronounced Raman signal enhancement imputable to plasmonic effects.

Keywords: Pd; Pt; FTO; laser irradiations; dewetting; nanoparticles

---

1. Introduction

Palladium (Pd) and Platinum (Pt) nanoparticles (NPs) are widely exploited to design and fabricate nanostructured innovative sensing, catalyst, electronic, energy production and storage devices [1–8]. From a general point of view, this type of nano- system captures great interest due to size-dependent structure and properties, high specific surface area and reactivity. From a specific point of view, for example, some notable chemical reactions occurring at surfaces take advantage of the enhanced catalytic activity of Pd and Pt when in the nanoscale form [9–13]. In addition, the physico-chemical properties of Pd and Pt NPs can be largely tuned by the control of their size and shape. This is an important property of the NPs for several technological applications. A critical issue is the formation of NPs with a desired size that should take place directly on a support. Hence, for the exploitation of these nanostructures in real devices, the design of simple, cost-effective, versatile, high-throughput fabrication methods on suitable supporting surfaces, allowing desired NPs size and shape control, is of paramount importance [14]. Obviously, the detailed understanding of the basic microscopic mechanisms governing the process involved is crucial in assuring the desired NPs control.

So, we report here on the exploitation of a laser-based approach to fabricate Pd and Pt NPs supported on a functional surface. The approach is based on the sputtering-deposition of nanoscale-thick Pd or Pt films onto the surface of a Fluorine-doped Tin Oxide (SnO_2:F) layer on glass (soda-lime), that is, Fluorine-doped tin oxide (FTO/glass) substrate. FTO is a transparent conductive oxide which use is largely diffused as coating in solar cell devices or for photoelectrochemical reactions on surface [15–19]. Then, nanosecond laser irradiations were carried on the metallic films surface so to exploit the potentialities of the laser-matter interaction for surface nanostructuration [20–30]. In particular, we processed, by the laser pulse, Pd or Pt films of various thickness d [20–37]. The molten-phase dewetting process of the Pd and Pt films is observed to lead to the formation of NPs for which the mean diameter $<D>$ and mean surface density $<N>$ are quantified. Our results show, in particular, the effect of the FTO surface topography on the resulting NPs size through the geometry-dependent chemical potential, an effect which is absent for the dewetting phenomenon of thin metal films on flat surfaces [22–32]. This effect is discussed starting from previous results on metallic film dewetting occurring on intentionally patterned surfaces [33–36]. We show that this effect is crucial in the present case since the used FTO substrate (similarly to standard FTO used in real devices) does not present a flat surface on a microscopic scale. Instead, due to standard deposition processes [17,38,39], the FTO layer on the glass slide surface is obtained as structured in micrometric randomly-arranged FTO pyramids.

Finally, we present, also, Raman analysis on the fabricated samples finding, for the case of the Pd NPs, a pronounced Raman signal enhancement imputable to intense plasmonic effects.

From a general point of view, this work completes our previous studies on the molten phase dewetting of pure (Au) or alloy (AuPd) metals films on transparent conductive oxides [17,40,41], extending these studies to other metals such as Pd and Pt whose nanostructuration by laser pulsed-induced dewetting is rarely studied (Pt) [42,43] or, practically, absent (Pd).

2. Materials and Methods

Fluorine-doped Tin Oxide (SnO_2:F, FTO)/glass (soda-lime) slides, from KINTEC factory (Hung Hom, Kowloon, Hong Kong) [44], were the starting substrates. According to our previous characterizations [17], the average transmittance of these slides, corresponding to incident light wavelength from 400 to 1100 nm, is 73.2% and the FTO sheet resistivity is 8.6 ohm/sq. For each metal deposition, some glass/FTO slides were inserted within the vacuum chamber of a RF Emitech K550X sputter coater (Quorum Technologies, East Sussex, UK). In addition, Si slides were introduced with the glass/FTO slides. They served as reference samples for successive check of the effective thickness of the deposited metal films. Pd or Pt films were sputter-deposited onto the FTO surface using 99.999% purity Pd or Pt targets using Ar as sputtering gas. In particular, the control of the effective thickness d of the deposited metal films is reached by the control of the deposition time (in the range 1–240 s) and the emission current (in the range 5–50 mA). Successive Rutherford backscattering spectrometry analysis (using 2-MeV $^4He^+$ incident ions and backscattered at 165°) on the metal films deposited on the reference Si slides allowed the association of several couples (deposition time, emission current) to a specific thickness d of the Pd or Pt film. For the analysis presented in this work we chose, in particular, d_{Pd} = 3.0, 7.5, 17.6, 27.9 nm for the Pd films, and d_{Pt} = 1.9, 7.5, 12.2, 19.5 nm for the Pt films, with a measurement error of 5%.

The metal films supported on the FTO surface of the glass/FTO slides were processed by nanosecond laser irradiation (one pulse per film). In order to carry out the irradiations, we used a nanosecond pulsed (12 ns) Nd: yttrium aluminum garnet YAG laser working at 532 nm (Quanta-ray PRO-Series pulsed Nd:YAG laser, Spectra Physics, Santa Clara, CA, USA). In our experimental setup, the laser spot is circular with a diameter of approximately 4 mm. A Gaussian shape characterizes the spatial distribution of the laser intensity. Specifically, the laser intensity decreases from the chosen maximum value to the 97% of this maximum value within a circular area of 600 μm in diameter. For the experiments presented in this work, a laser fluence having as maximum 0.50 J/cm^2 (measurement error of 0.025 J/cm^2). So, we consider that within a circular spot with diameter 600 μm centered on the entire laser spot (4 mm in diameter), the metal film was irradiated by one pulse with 12 ns duration and fluence 0.50 J/cm^2. All the analyses refer, then, to the NPs formed in this region.

Microscopic morphological analyses were carried out by using a scanning electron microscope (SEM, Zeiss FEG-SEM Supra 25 Microscope, Carl Zeiss Microscopy, New York, NY, USA). The Gatan Digital Micrograph software (version 3.0, GATAN Inc., Pleasanton, CA, USA) was used to extract quantitative information by the SEM images. In particular, to extract from the SEM images the mean diameter <*D*> of the metal NPs, for each image a threshold on the brightness of the image was set, so that the bright regions in the image, with intensity value 1, represent the NPs and the dark regions, with intensity value 0, represent the supporting substrate. The diameter *D* of a NP is measured as the diameter of the smaller circle inscribing the NP. The mean value <*D*> of the diameter of the NPs (and the corresponding error arising as the standard deviation) for each sample has been quantified on a statistical population of 400 NPs. In addition, the mean surface density <*N*> (number of particles per unit area) was evaluated by direct counting and averaging on several SEM images.

Raman spectra on the samples were acquired by using a micro-Raman spectrometer (Confocal Raman–AFM SNOM, WITec ALPHA300RS, Ulm, Germany) equipped with a charge-coupled device (CCD) system. A 532 nm laser line (output power ~32 mW, power on the sample ranging from 0.5 to 10 mW), a grating with 600 grooves mm^{-1} and a 100× differential interference contrast (DIC) objective (NA 0.9) were used.

3. Results and Discussion

First, we analyzed, by SEM imaging, the topography of the reference FTO surface. So, Figure 1 reports a typical SEM image of the bare FTO surface. It appears that the FTO layer on the glass substrate has a complex non-flat topography. In fact, by the inspection of this plan-view image, the FTO layer results to be formed by micrometric structures having an approximately truncated-pyramidal shape and which, randomly arranged, cover the entire surface of the glass slide. We point out that this is the standard morphology for deposited FTO layers [38,39] and that it is desired in several applications (solar cells, etc.) due to its ability in efficiently scatter light [17,18,39,41]. Then, in Figures 2 and 3, some exemplificative microscopies of the FTO surface covered by the Pd and Pt films are reported. Figure 2 refers to the Pd covered FTO surface with increasing effective thickness of the covering metal film: d_{Pd} = 3.0 nm (a) and d_{Pd} = 27.9 nm (b). Figure 3 refers to the Pt covered FTO surface with increasing effective thickness of the covering metal film: d_{Pt} =1.9 nm (a) and d_{Pt} = 19.5 nm (b). Increasing the amount of sputtered Pd or Pt (i.e., increasing the effective thickness of the deposited metal film), an evolution of the surface morphology can be observed: the pyramidal structuration of the FTO layer is yet recognizable; however, over the pyramids' surfaces a nano-granular morphology develops (see, in particular, Figures 2b and 3b)

Figure 1. FTO bare surface as imaged by SEM. The surface structuration in pyramidal-like structures is recognizable.

This morphology is the standard one, in the initial stages of growth (i.e., nucleation and growth) for metal films deposited on non-metal substrates and which leading growth process is the Volmer-Weber growth mode [3,17,25,26,31,41,45–47].

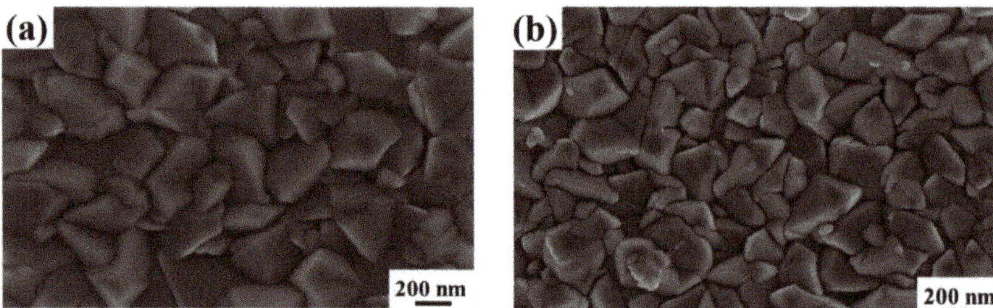

Figure 2. SEM images of the FTO surface covered by Pd film with effective thickness of: (**a**) 3 nm; (**b**) 27.9 nm. The presence of the films can be recognized in more rough pyramids surfaces with respect to bare surfaces.

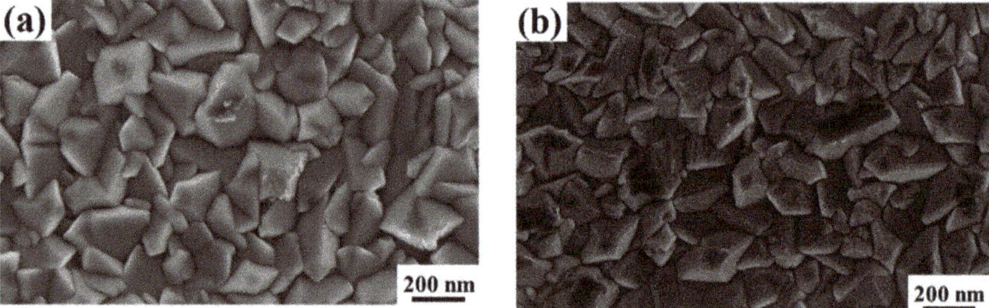

Figure 3. SEM images of the FTO surface covered by Pt film with effective thickness of: (**a**) 1.9 nm, (**b**) 19.5 nm. The presence of the films can be recognized in more rough pyramids surfaces with respect to bare surfaces.

However, in late stages of growth, increasing the amount of deposited atoms, the nucleated small metal islands grow until to contacting each other giving rise to a coalescing process [45–47]. In this stage, a percolative surface morphology of the metal film is obtained in the sense that it is formed by ramified interconnected nano-islands separated by small gaps [45–47]. Finally, continuing to deposit more and more material, holes in the growing metal film are filled to form a continuous rough film [45–47].

On each metal-covered sample, we carried out the laser irradiation. Then, we used the SEM analysis to study the metal surface morphology evolution versus the metal film thickness, see Figures 4 and 5. In general, the pulsed laser-induced dewetting of the metal film is observed, that is, the 0.50 J/cm^2 laser irradiation for 12 ns causes the melting of the Pd or Pt film followed by the film rupture and retreating processes towards the formation of NPs of circular section (Figures 4 and 5). The spherical (or almost-spherical) shape of the NPs can be easily recognized from the SEM images in Figures 4 and 5 where in several cases NPs are observed from a tilted configuration due to the surface structuration of the FTO surface (i.e., the NPs placed on the lateral surfaces of the FTO pyramids). In fact, generally, metal films on transparent conductive oxides are highly non-wetting systems [26] so that, after the dewetting process, spherical or almost-spherical NPs are obtained (i.e., maximization of the contact angle).

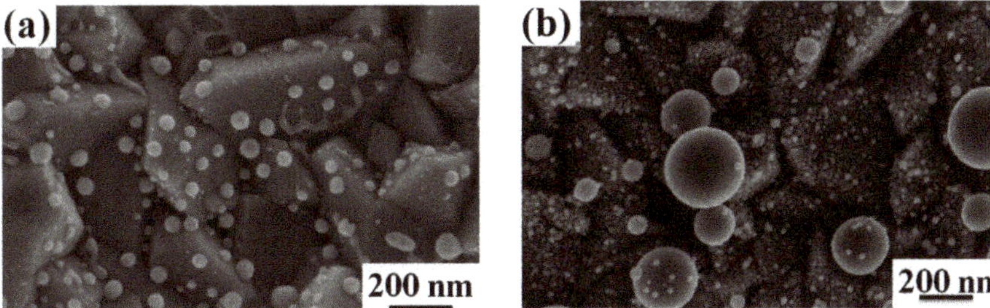

Figure 4. SEM images of the FTO surface covered by Pd film after the 0.50 J/cm^2 laser pulse ((**a**) 3 nm-thick, (**b**) 27.9 nm-thick. The formation of almost-spherical particles on the FTO surface is recognizable both in (**a,b**).

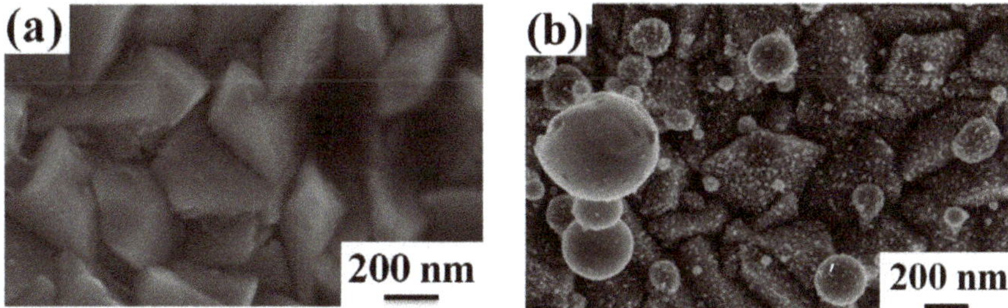

Figure 5. SEM images of the FTO surface covered by Pt film after the 0.50 J/cm^2 laser pulse ((**a**) 1.9 nm-thick; (**b**) 19.5 nm-thick). In (**a**) the Pt film is unchanged, while in (**b**) the formation of almost-spherical particles on the FTO surface is recognizable.

In particular, Figure 4 presents typical SEM microscopies showing the Pd NPs originated from the pulsed laser irradiation of the FTO-supported Pd film having thickness (a) 3.0 nm or (b) 27.9 nm; in addition, Figure 5 presents SEM microsopies showing the Pt NPs originated from the irradiation of the FTO-supported Pt film with thickness (a) 1.9 nm (no NPs formation) and (b) 19.5 nm.

It is, nowadays, well-established [17,22–27,41–43] that during the nanosecond pulsed laser irradiation of deposited nanoscale-thick metal films, the NPs are formed from the continuous films by a molten-phase dewetting process. In fact, in standard metals (such as Pd and Pt) the thermal equilibrium between hot electrons and phonons is established, typically, within a characteristic time $t_{eq} \approx 50$ ps [17,22–27,41–43,48]. Using nanosecond pulsed laser to generate heat in the film, as in the present experiments for which the pulse duration is $\tau = 12$ ns so that $\tau >> t_{eq}$, then the metal film melting dynamics is the main process [17,22–27,41–43,48]. Therefore, in the present experiments, the Pd and Pt dewetting process occurs with the films being in the molten state (i.e., spinodal dewetting [17,22–27,41–43,48]) followed by the film's solidification. Shortly, it is widely accepted that this dewetting process starts by the film perforation which leads to holes formation in the film. These perforations are likely to occur where the film is thinner since the fluence required for melting is lower for thinner films [24]. Deposited metal films, in fact, are, typically, not perfectly uniform in thickness presenting, instead, a natural roughness (local variation in thickness). The metal at the hole edges forms the so-called rims which retract drawing away from the center of the hole. The retreating molten films in neighboring holes coalesce into metal filaments (some solidified Pd filaments are recognizable, for example, in Figure 4a). These filaments are thermodynamically unstable so that, then, then split into droplets by the Rayleigh instability [24–27] and minimizing the total surface energy of the system. In the final stage, after solidification, an array of metal NPs uniformly covering the laser-irradiated surface is obtained.

We proceeded, then, to the quantitative analysis of the Pd and Pt NPs characteristics using the SEM images. In particular, for each sample, distributions of the NPs diameter D were constructed as described in the experimental section. As an example of the procedure, Figure 6a reports a part of the SEM image of Figure 4a concerning the NPs obtained by the laser irradiation of the 3 nm-thick Pd film and in Figure 6b how it was treated in luminosity and contrast to identify the NPs (circular white regions). Using the scale marker, for each circular region the diameter can be identified so to construct the NPs diameters distribution. In addition, by direct counting of the circular white regions, the NPs surface density can be derived.

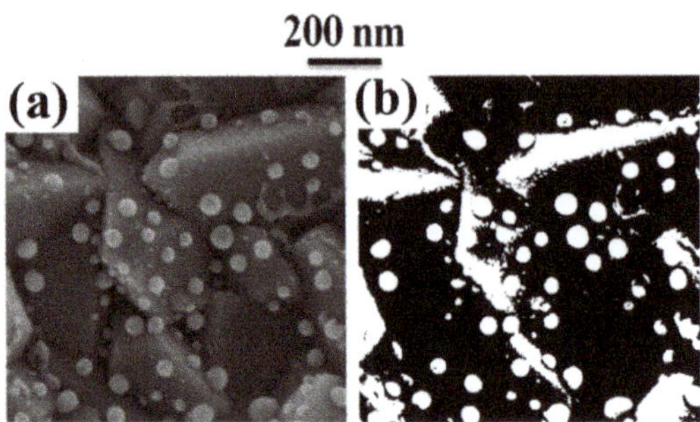

Figure 6. (**a**) SEM image showing the NPs produced on the FTO surface after the laser irradiation of the 3 nm-thick deposited Pd film (it is a part of the SEM image in Figure 4a). (**b**) SEM image in (**a**) suitably treated in luminosity and contrast to identify the NPs (circular white regions) so to extract NPs size and number per unit area.

Figure 7 reports, then, some examples of obtained diameter distributions: (a,b) are the diameters for the NPs obtained by the laser irradiation of the 3 nm-thick ((a)) and 27.9 nm-thick ((b)) Pd films deposited on the FTO surface; (c,d) are the diameter distributions for the NPs obtained by the laser irradiation of the 7.5 nm-thick ((c)) and 19.5 nm-thick ((d)) Pt films deposited on the FTO surface. As easily recognizable, (a,c) shows a monomodal size distribution while (b,d) bimodal sizes distributions.

Each diameter's distribution was, hence, statistically analyzed to extract the mean diameter $<D>$ (and the corresponding standard deviation ΔD) of the NPs. In addition, we evaluated, also, the mean NPs surface density $<N>$ and the corresponding errors. The results are summarized in the plots in Figure 8. In particular, performing these analyses, in some samples we observed a monomodal size distribution (one gaussian distribution) indicative of an unique NPs population to which was associated an uniqhe mean diameter $<D>$ and mean NPs surface density $<N>$. In other samples, we observed a bimodal size ditribution (two gaussian distributions) indicative of two NPs sub-populations to which were associated mean diameters $<D_1>$ and $<D_2>$ and mean NPs surface densities $<N_1>$ and $<N_2>$.

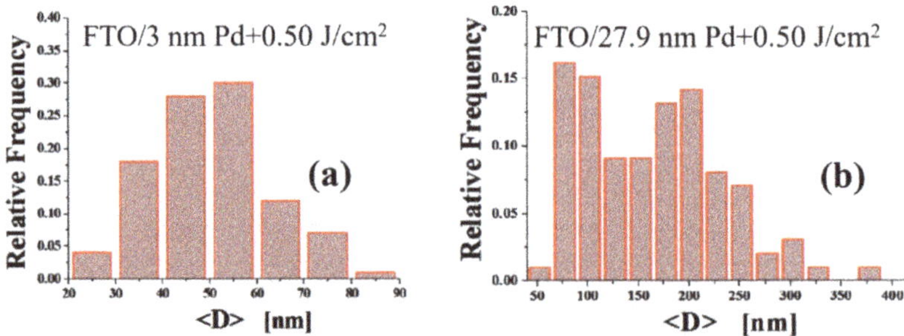

Figure 7. *Cont.*

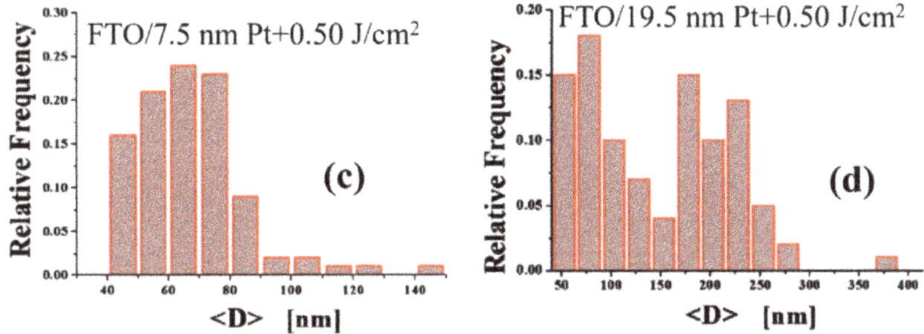

Figure 7. Examples of NPs diameter distributions: The diameters distributions for the NPs obtained by the laser irradiation of the (**a**) 3 nm-thick and (**b**) 27.9 nm-thick Pd films deposited on the FTO surface; The diameters distributions for the NPs obtained by the laser irradiation of the (**c**) 7.5 nm-thick and (**d**) 19.5 nm-thick Pt films deposited on the FTO surface. As easily recognizable, (**a,c**) shows a monomodal size distribution while (**b,d**) bimodal sizes distributions.

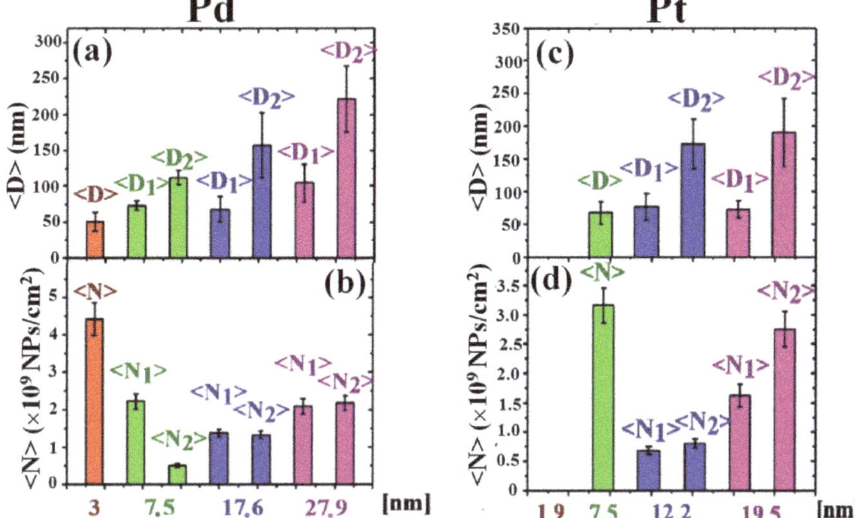

Figure 8. Mean diameter $<D>$ ((**a**)) and mean surface density $<N>$ ((**b**)) of the Pd NPs obtained by the 0.50 J/cm^2 laser irradiation of the 3 nm, 7.5 nm, 17.6 nm, 27.9 nm thick deposited Pd films. Mean diameter $<D>$ ((**c**)) and mean surface density $<N>$ ((**d**)) of the Pt NPs obtained by the 0.50 J/cm^2 laser irradiation of the 1.9, 7.5, 12.2, 19.5 nm-thick deposited Pt films. The indication of two values for $<D>$ or $<N>$ for the same sample means the presence of two different NPs sub-populations in that sample.

Specific values for $<D_1>$, $<D_2>$, $<N_1>$ and $<N_2>$ are, further, summarized in Table 1 for the various samples.

To discuss these data, first of all we recall some basis of the thin metal film melting process induced by pulsed laser irradiations. In the first approximation, a simple model [24,26] considers the metal film of thickness d as irradiated by one laser pulse of duration τ with a top-hat temporal profile and establishes (as reliable hypothesis) the total heat (per unit area) generated in the film as $S = I\tau(1-r)[1-\exp(-\alpha d)]$ being I the laser pulse intensity (W/cm^2), r the film reflectivity at the laser wavelength, α^{-1} the film optical absorption length at the laser wavelength (α is called the absorption coefficient). Note, however, that for a thin film, r and α depend on d. On the basis of this model, the consequent temperature rise in the thin film is $\Delta T = S/Cd$ where C is the film heat capacity per unit volume. So, the melting (and molten-phase dewetting) characteristics of a thin metal film depend on optical and thermal characteristics of the film (d, r, α, C, melting temperature). In principle, differences in these parameters for Pd and Pt justify the observed differences in $<D>$ and $<N>$ when fixed the film thickness, the laser fluence and pulse duration. This is the case for the 7.5 nm-thick Pd and Pt films for which, on the same substrate, two Pd NPs populations and an unique Pt NPs population are, respectively, obtained. On the other hand, it is easy to understand why the laser irradiation of the 1.9 nm-thick Pt film does not produce a dewetting process (i.e., formation of the NPs). In this case, clearly, $d<<\alpha^{-1}$ (for metal films, generally, $\alpha^{-1}\sim 10$ nm) and/or r is so low that the Pt film does not absorb energy from the laser pulse: the film temperature does not change and the film is unaffected by the laser irradiation.

Table 1. Summary of the values of $<D_1>$, $<D_2>$, $<N_1>$, $<N_2>$ for the NPs in the samples.

Thickness of the Film-Material	$<D_1>$ (nm)	$<N_1>$ ($\times 10^9$ cm^{-2})	$<D_2>$ (nm)	$<N_2>$ ($\times 10^9$ cm^{-2})
3 nm-Pd	~50	~4.5	0	0
7.5 nm-Pd	~75	~2	~100	~0.5
17.6 nm-Pd	~70	~1.5	~150	~1.5
27.9 nm-Pd	~100	~2	~220	~2
1.9 nm-Pt	0	0	0	0
7.5 nm-Pt	~70	~3.2	0	0
12.2 nm-Pt	~75	~0.6	~220	~0.6
19.5 nm-Pt	~70	~1.6	~180	~2.7

Shortly, we also discuss the laser irradiation effect on the underlaying FTO substrate. In fact, the melting temperatures for Pt and Pd are, about, 2040 and 1828 K, respectively, while the melting temperature for the SnO_2 is around 1900 K. So, in the laser irradiation processes we could expect, also, a partial melting of the FTO substrate due to the high temperatures achieved for melting, at least, the Pt films. To discuss this point, for example, we recall the results obtained by Font et al. [49] regarding the heat transfer mechanisms involved in thin metal films melting on a SiO_2 layer and exposed to laser irradiation. In this work, the authors present a model describing the heat transfer and phase change of the metal/substrate system during nanosecond laser irradiation with parameters similar to those presented in our work. In particula, Font et al. [49] considered a Cu film/10 nm-thick SiO_2 layer/Si substrate system irradiated by a pulsed laser characterized by a Gaussian beam, and they studied the effect of the laser parameters, including energy density and pulse duration, on possible melting of the SiO_2 layer. Interestingly, they found that maximum thickness of the melted SiO_2 region can be of the same order of magnitude as the thickness of the metal film. The calculations carried out by the authors are based on some assumptions. In particular, for example, they supposed that the substrate is completely transparent to the laser radiation, so that the metal film absorbs the energy of the laser and transfers it to the substrate by conduction. This condition is fulfilled, also, by the FTO/glass substrate used for the experiments presented by us. To describe the heating and possible melting of the substrate, they considered, obviously, heat conduction in each layer of the system. In the basing equations, so, the thermal conductivity of the SiO_2 layer (in solid and liquid phases) is involved. This value, however, is very different from that characterizing the FTO layer. In fact, for example, in the solid phase, for SiO_2 typical value for the thermal conductivity is 1.4 W/m·K [49]. Instead, SnO_2 is characterized by a much higher thermal conductivity, of the order of 40 W/m·K [50]. Clearly, the temperature increase in the film depends on how it dissipates the laser-generated heat [24]: concerning a metal film deposited on a low thermal conductive substrate, the laser-generated heat will remain, mostly, confined in the metal film resulting in a higher increase of temperature of the system. This means that for the metal films deposited on the FTO substrate the rapid dissipation of the laser-generated heat through the FTO substrate results in a lower increase of the system temperatrure compared to the case for which the same metal films are deposited on a lower thermal conductivity substrate (as SiO_2). In fact, comparing the SEM images in Figures 4 and 5 (Pd and Pt films on FTO after the laser irradiations) to the SEM image in Figure 1 (bare and untreated FTO), it seems clear that the surface morphology of the FTO substrate is unchanged after the laser irradiations. This is an indication of the fact that even if the laser-generated heat determines the temperature increase of the metal films above the film material melting temperature, the relatively high thermal conductivity of the FTO probably results in a temperature increase of the FTO layer below its melting temperature. This is the situation for the 0.50 J/cm^2 laser pulse. This situation is strongly changed, for example, using a 0.75 J/cm^2 laser pulse for which the SEM images (not shown) clearly showed a dramatic change of the FTO surface morphology: in this case, the FTO pyramids are no more recognizable being the FTO surface more similar to a flat surface. This is, probably, consequence of the higher temperature reached by the FTO layer, higher than the FTO melting temperature. In this case, so, also the FTO layer melts during the laser irradiations and, also, an FTO morphology change is involved in the overall process. This is a further reason for which we limit the data presented here to the 0.50 J/cm^2 laser pulse, so involving only the Pd and Pt molten-phase dewetting process.

Regarding the standard dewetting process of metal films on flat surfaces [22–31], a unique NPs population is expected with a mean size, which increases by increasing the starting thickness of the deposited film. The observation of the formation of two different NP populations in some samples is, thus, indicative of the effect of the substrate non-flat surface topography on the film dewetting process, influencing, then, the size of the NPs. To describe this effect we, briefly, discuss some of the results due to Gierman and Thompson [33]. They intentionally patterned, by using electron beam litography, a SiO_2 surface in spatially ordered arrays of inverted pyramid shaped pits. However, different patterns were designed by changing the spatial period of the pits (175 or 377 nm) and the pit-to-mesa width ratio (1.5, 1.9, 5.3). Onto the so-patterned SiO_2 surface, Au films of different thickness (16, 21, 32 nm) were evaporated. The Au films solid-state dewetting process was activated by 800 °C annealing process. The microscopic analysis revealed for the resulting Au NPs originated in the dewetting process, a spatial arrangement on the substrate surface and size distribution dependent on the geometric characteristics of the surface pattern (spatial period of the pits and pit-to-mesa width ratio) in combination with the Au film thickness. On the basis of the combination of these parameters: in some cases the preferential formation of a single NP per pits with no NPs on the mesa was observed; in other cases, the formation of one NP per pit with some other NPs on the mesa was achieved; in further cases, the generation of multiple NPs per pit with some other NPs on the mesa resulted; finally, some other combinations lead to the formation of large NPs on the mesa (often covering the pits) with no NPs in the pits. These results clearly demonstrated the crucial effect of the surface topography on the deposited metal film dewetting also giving the possibility to study, quantitatively, this effect taking into in account the excess of local chemical potential as modulated by the finite local curvatures on the substrate surface [33]. In fact, considering the dewetting process of a thin metal film on a flat surface, it is clear that the driving force for the process is, solely, the minimization of the total surface and interface energy of the system. For a flat surface, the local radius of curvature is, in each surface point, $R \rightarrow \infty$. The corresponding local curvature is $\kappa = 1/R$ to which is related, by the Gibbs–Thomson relation, the local excess of chemical potential $\Delta\mu = \kappa\gamma\Omega \rightarrow 0$ (being γ the film surface energy and Ω film atomic volume. Consequently, the dewetting process of a thin film on a flat surface is not affected by the surface topography. The situation changes if the surface is not flat, that is, if it presents finite local curvatures (as in the case of the patterned surfaces used by Giermann and Thompson). In this case, due to the finite values of R, the non-zero values of $\Delta\mu = \kappa\gamma\Omega$ introduce an additional driving force for the film dewetting process determining, in particular, a preferential material diffusion from the positions with $\kappa > 0$ (peaks or ridges) to the positions with $\kappa < 0$ (valleys), see the representative scheme in Figure 9. It is clear, then, the modulation of $\Delta\mu$ through the surface geometry impacts on the dewetting process and on the resulting characteristics (spatial arrangement and size) of the NPs originanting from this process. Refererring to the specific case reported by Giermann and Thompson [33], which is particularly useful for the interpretation of our data, the edge of a pit is characterized by a $\Delta\mu \propto 1/R_A > 0$ while the apex of an inverted pyramid is characterized by $\Delta\mu \propto -1/|R_B| < 0$, see Figure 9. It follows that, to minimize the total surface and interface energy of the system and to establish a global condition for which $\Delta\mu = 0$, the film material will diffuse away from the pit edge toward the pit apex and here, preferentially, will form the NPs. However, since the local curvature at the pit edge and the apex decreases with increasing film thickness, the driving force for flow from the edge to the apex also decreases with increasing film thickness. So, overall, the effect becomes dependent on the film thickness.

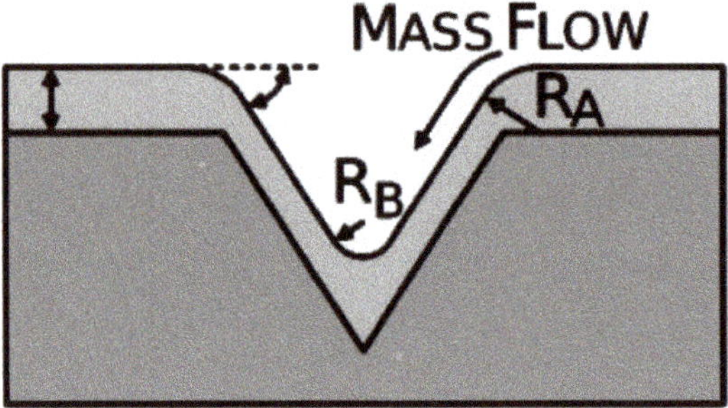

Figure 9. Representation of a thin film conformally covering a patterned surface. The curvature at the pit edge, R_A, and at the inverted apex, R_B are shown. Reprinted with permission from [33]. Copyright 2005 AIP Publishing.

In our case, we observed that increasing the thickness of the Pd film from 3.5 to 27.9 nm and of the Pt film from 7.5 to 19.5 nm, the NPs population evolves from monomodal to bimodal. This indicates the existence of a critical film thickness d_c identifying a variation in the dewetting characteristics of the film, as pictured by the combination of Figures 10 and 11. For $d < d_c$ (generally named "low-thickness" condition) the substrate topography does not affect the film dewetting process (Figure 10). In this condition, the film does not interact with the topographic features of the substrate and, then, the substrate behaves as a flat substrate with respect to the film dewetting. Hence, the film dewets as on a planar substrate originating NPs uniformly arranged over the surface of the substrate. In addition, the size of the dewetted NPs, being the starting thickness of the deposited film very low with respect to local surface curvatures (defined by the characteristic sizes of the FTO pyramids, i.e., height, width, spacing), is not influenced by the substrate topography. For $d > d_c$ (generally named "high-thickness" condition) the geometrical features of the substrate affect the film dewetting process and, hence, the resulting mean size of the formed NPs (Figure 11). In this case the thickness of the deposited film is, at least, comparable to the local surface curvatures. Then, the film dewetting is driven, in addition to the surface energy minimization, by the surface topography through the local excess of chemical potential. This results in the preferential formation of the NPs in-between or over the pyramids. It is clear, hence, that the characteristic sizes of the FTO pyramids affects the final NPs mean size. The overall effect is determined by an interplay between these characteristic sizes (establishing the local surface curvatures) and the thickness d of the film. Therefore, observing Figure 8, in that samples where an unique population of NPs is obtained, the situation $d < d_c$ is realized (dewetting as on a flat surface), whereas in that samples where a two NPs sub-populations are obtained, the situation $d > d_c$ is realized (substrate topography-driven dewetting). Then, we can conclude, also, that 3 nm $< d_c <$ 7.5 nm for the Pd film and that 7.5 nm $< d_c <$ 12.2 nm for the Pt film. However, for sake of completeness, we also mention that, according to previous experimental and theoretical studies on dewetting of films on patterned surfaces [51–54], a critical thickness d_c for the film separating different effects of the substrate topography on the dewetting process is related to additional parameters than the local excess of chemical potential $\Delta\mu$. In particular, it is known that a fixed substrate pattern can influence the dewetting pathways of supported film depending on the film thickness by acting on the length-scale of the processes (holes nucleation, film retreating, Rayleigh instability, etc.) involved in the dewetting phenomenon. Dewetting on a flat substrate progresses with the formation and growth of randomly placed holes with a certain mean length-scale. Eventually, coalescence of holes produces an isotropic collection of droplets. The average diameters of the dewetted structures depend on the initial film thickness. On the other hand, specific spatial periodicity of patterns on the substrate can act on a deposited film by imposing conditions on the length-scale of the processes involved in the dewetting phenomenon (in addition to the effect imposed by local excess of chemical potential on the material film diffusive processes). An a-priori theoretical determination of the critical thickness d_c should take into in account, also, such effects but this is out of the scope of the present experiments.

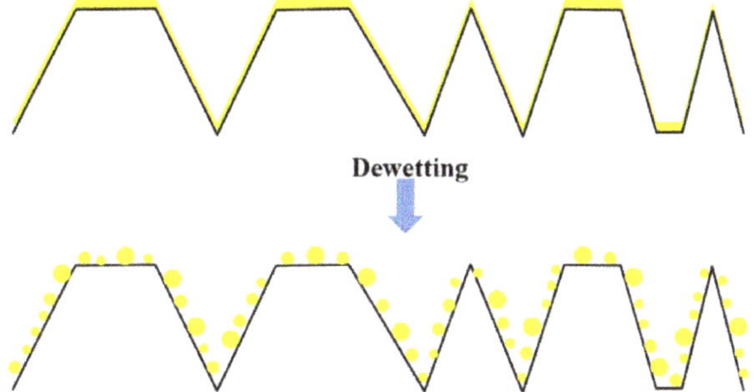

Figure 10. Picture of the result of the "low-thickness"-film dewetting process on the FTO topographically structured surface. Small NPs are uniformly distributed over the entire FTO surface (as on a flat surface).

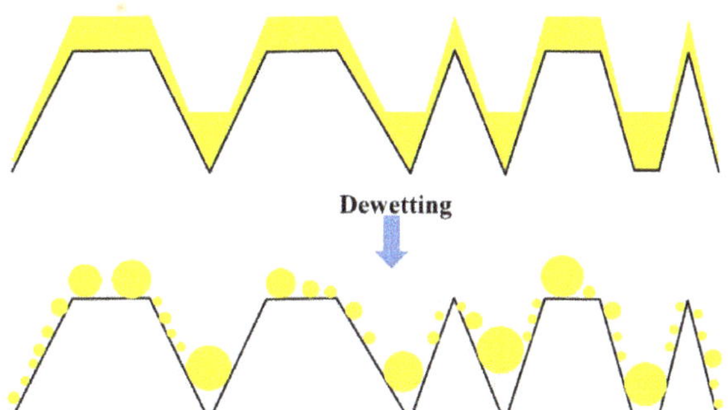

Figure 11. Picture of the result of the "high-thickness"-film dewetting process on the FTO topographically structured surface. Large NPs are preferentially formed in-between or over the FTO pyramids.

To conclude the analyses of the characteristics of the Pd and Pt NPs produced on the FTO substrate, we performed, also, Raman analysis. Figure 12 displays, first of all, the Raman spectrum of the bare FTO/glass substrate, in black. There are two fundamental Raman scattering peaks which are those characteristic of rutile SnO_2 single crystal [55]. For pure SnO_2, the characteristic and intense band at 625 cm^{-1} is due to the A_{1g} vibration mode of SnO_2. The weak band around 478 cm^{-1} is due to the E_g vibration modes of SnO_2 [55]. Figure 12 reports, also, as examples of the general behaviour, Raman spectra of the FTO/glass substrate covered by Pd or Pt NPs originating from the laser heating of the deposited films: in red the spectrum of the substrate covered by Pd NPs obtained by the laser irradiations of the 27.9 nm-thick Pd film; in blue the spectrum of the substrate covered by Pd NPs obtained by the laser irradiations of the 17.6 nm-thick Pd film, in green the spectrum of the substrate covered by Pt NPs obtained by the laser irradiations of the 19.5 nm-thick Pt film. The general effect is that, compared with bare FTO, the intensity of Pd NPs/FTO samples become stronger.

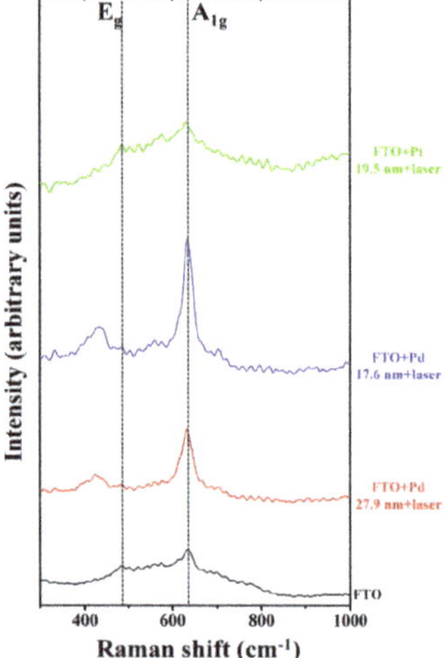

Figure 12. Raman spectra corresponding to the bare FTO substrate (black), FTO covered by Pd NPs obtained by the laser irradiation of the 27.9 nm-thick Pd film (red) and of the 17.6 nm-thick Pd film (blue), FTO covered by Pt NPs obtained by the laser irradiation of the 19.5 nm-thick Pt film. SERS effect can be particularly recognized in the blue spectrum.

On the contrary, the presence of the Pt NPs on the FTO surface, does not influence the FTO peak intensity, indicating no Pt NPs-FTO interaction. In particular, the Pd NPs SERS (Surface Enhancement Raman Scattering) enhancement may be due to Pd NPs surface plasmon resonance [5,56,57]. However, surface plasmon resonance effects also for Pt NPs are expected [5]. The reasons for which these effects lead to intense SERS in the Pd NPs/FTO samples while no SERS effects in the Pt NPs/FTO samples is actually under investigation as a perspective work, requiring a crossing of experimental and simulation-based results. Here, we limit to consider that the SERS enhancement should arise from the effective resonance coupling between the localized surface plasmon of the NPs and the characteristic parameters of the energy band structure of the FTO. A match should be realized in the case of the Pd NPs and the FTO, contrary to the case of the Pt NPs and FTO. Besides, several other factors, such as the Ohmic loss, non-radiative Forster energy transfer, lower surface plasmon radiative efficiency, may be responsible for the absence of SERS enhancement in the Pt NPs-FTO systems and these effects have to be considered in a future complete description. However, we mention, for example, that the optical loss of NPs material can be used as a measure to predict the effect in SERS, whereas the optical loss is intended as the imaginary part (ε_2) of the dielectric constant $\varepsilon(\omega) = \varepsilon_1(\omega) + i\varepsilon_2(\omega)$ [58]. In fact, the imaginary part of the dielectric constant causes Ohmic damping of the electrons oscillations in the metal. Our SERS data seem in agreement with the consideration that several experimental and theoretical works found higher values of ε_2 for bulk, thin films, NPs materials of Pt than of Pd in the UV range [59–61].

We observe, finally, in Figure 12, a shift (toward lower wavenumbers) of the E_g peaks in the Pd NPs/FTO samples with respect to the bare FTO substrate, while this shift is absent in the Pt NPs/FTO samples. It is absent, also, before the laser treatment so that it cannot related to substrate modifications due to the sputtering deposition processes. From a general point of view, in Raman spectra, shifting of peaks towards lower or higher wavenumber is related to several factors [62] such as change in the chemical bond length within the sample (the shorter bond length causes to shift higher wavenumber or vice versa), changes in temperature and stress within the sample, and so on. More specifically, a shift in

Raman peaks could be due to the addition of oxygen in the sample. In this regard, the fact that this shift is present only in the case of the Pd NPs could suggest its relation to the Pd NPs surface oxidation [63]. It is widely recognized that Pd NPs are subjected to the interaction with environmental oxygen to form a PdO surface shell [64], while Pt NPs are more stable against this reaction [65]. We can support the conclusion that the Raman peak shift for the sample with Pd NPs could be related to surface oxidation of the NPs, by considering that previous works reported wavelength shift of plasmonic bands for Pd NPs [66] and Ag NPs [67,68] after exposure to air or oxygen with consequent formation of surface oxide (PdO or Ag_2O) for the NPs. To conclude, we recall that the Pd and Pt NPs produced have a quasi spherical shape and we comment on this fact in view of the Raman measurements results: Raman scattering, which is based on inelastic light scattering upon vibrational excitation of molecules and materials is an extremely inefficient process with about one photon out of 10^7 being inelastically scattered [69]. This drawback can be overcome when the molecules are located near a rough metal surface or metal NPs, which results in a boost of Raman scattering, in the process known as SERS. Upon resonant excitation of strong localized surface plasmon resonances, a greatly enhanced local electromagnetic field (near-field) is generated at the NP surface. This near-field in turn couples to the incident field and re-reemits radiation with the same wavelength (resonant scattering). As the scattering cross section scales with the square of the polarizability (the degree on how easily the electron cloud can be displaced) also the near-field strongly increases for anisotropic NPs due to higher polarizability, as compared to spheres. So, elongated NPs or those with edges and corners, such as nanorods, nanocubes, and nanotriangles, thus generate strong near-fields concentrated at vertices. The presence of the plasmonic near field at the NP surface upon plasmonic excitation increases the polarizability of an adsorbed molecule (compared to a free molecule) due to mutual excitation between induced dipole in the molecule and induced dipole in the NP and leading to enhanced Raman scattering. On the basis of these considerations, a further perspective on the present work lies in the establishment of laser process parameters to controllably produce complex-morphology Pd and Pt NPs to reach higher SERS enhancements. This could be reached, for example, by using femtosecond pulsed laser irradiations [20,21].

4. Conclusions

Pd and Pt NPs onto non-flat FTO substrate were fabricated by exploiting the molten-phase dewetting process of deposited films. The Pd and Pt films were sputter-deposited on the FTO surface, this surface being constituted this surface by FTO microscopic pyramids randomly distributed over a glass slide. The dewetting process of the liquid metal films was initiated by nanosecond laser irradiations at 0.5 J/cm^2, resulting, after solidification, in the formation of Pd or Pt NPs onto the laser-irradiated FTO surface. However, analyzing the mean diameter *<D>* of the formed NPs versus the thickness *d* of the deposited films, an effect of the substrate topography was found. This effect was discussed by the description of the substrate topography influence on the excess of chemical potential driving the dewetting process. We observe that increasing the thickness of the Pd film from 3.5 to 27.9 nm and of the Pt film from 7.5 to 19.5 nm, the NPs population evolves from monomodal to bimodal. This indicates the existence of a critical film thickness d_c identifying a change in the dewetting characteristics of the film: for $d < d_c$ the substrate topography does not influence the film dewetting process, while for $d > d_c$ the topography of the substrate is crucial in determining the dewetting process and, so, the final mean size of the produced NPs. In that samples where an unique population of NPs is obtained, the situation $d < d_c$ is realized (dewetting as on a flat surface), whereas in that samples where a two NPs sub-populations are obtained, the situation $d > d_c$ is realized (substrate topography-driven dewetting). By analyzing the experimental results, we concluded that 3 nm $< d_c <$ 7.5 nm for the Pd film and that 7.5 nm $< d_c <$ 12.2 nm for the Pt film. Finally, the analysis of the Raman spectra for the bare FTO/glass sample and for the Pd NPs/FTO/glass samples and Pt NPs/FTO/glass samples indicated a SERS effect in the Pd NPs/FTO/glass samples due to the effective coupling between the localized surface plasmon of the Pd NPs and the FTO susbstrate while this effect is absent in the Pt NPs/FTO/glass samples.

Author Contributions: Conceptualization, F.R. and M.G.G.; Methodology, F.R., M.G.G., M.C., V.T.; Formal Analysis, F.R., M.G.G., M.C., V.T.; Investigation, F.R., M.G.G., M.C., V.T., G.P. and G.C.; Data Curation, F.R., M.C., V.T., G.P. and G.C.; Writing—Original Draft Preparation, F.R.; Writing—Review and Editing, F.R., M.G.G., M.C., V.T., G.P. and G.C.; Supervision, F.R., M.G.G.; Funding Acquisition, M.G.G.

Funding: This research received no external funding.

Conflicts of Interest: The authors declare no conflict of interest.

References

1. Feldheim, D.L.; Foss, C.A. *Metal Nanoparticles: Synthesis, Characterization, and Applications*; Marcel Dekker Inc.: New York, NY, USA, 2002.

2. Johnston, R.L.; Wilcoxon, J.P. *Metal Nanoparticles and Nanoalloys*; Elsevier: Amsterdam, The Netherlands, 2012.

3. Ruffino, F.; Crupi, I.; Irrera, A.; Grimaldi, M.G. Pd/Au/SiC nanostructured diodes for nanoelectronics: Room temperature electrical properties. *IEEE Trans. Nanotechnol.* **2010**, *9*, 414–421. [CrossRef]

4. Baca, M.; Cendrowski, K.; Kukulka, W.; Bazarko, G.; Moszyński, D.; Michalkiewicz, B.; Kalenczuk, R.J.; Zielinska, B. A Comparison of Hydrogen Storage in Pt, Pd and Pt/Pd Alloys Loaded Disordered Mesoporous Hollow Carbon Spheres. *Nanomaterials* **2018**, *8*, 639. [CrossRef] [PubMed]

5. Langhammer, C.; Yuan, Z.; Zorić, I.; Kasemo, B. Plasmonic Properties of Supported Pt and Pd Nanostructures. *Nano Lett.* **2006**, *6*, 833–838. [CrossRef] [PubMed]

6. Tran, M.; Whale, A.; Padalkar, S. Exploring the Efficacy of Platinum and Palladium Nanostructures for Organic Molecule Detection via Raman Spectroscopy. *Sensors* **2018**, *18*, 147. [CrossRef] [PubMed]

7. Liu, C.; Kuang, Q.; Xie, Z.; Zheng, L. The effect of noble metal (Au, Pd and Pt) nanoparticles on the gas sensing performance of SnO_2-based sensors: A case study on the {221} high-index faceted SnO_2 octahedra. *CrystEngComm* **2015**, *17*, 6308–6313. [CrossRef]

8. Kumar, M.; Bhati, V.S.; Ranwa, S.; Singh, J.; Kumar, M. Pd/ZnO nanorods based sensor for highly selective detection of extremely low concentration hydrogen. *Sci. Rep.* **2017**, *7*, 236. [CrossRef]

9. Vendelbo, S.B.; Elkjær, C.F.; Falsig, H.; Puspitasari, I.; Dona, P.; Mele, L.; Morana, B.; Nelissen, B.J.; van Rijn, R.; Creemer, J.F.; et al. Visualization of oscillatory behaviour of Pt nanoparticles catalysing CO oxidation. *Nat. Mater.* **2014**, *13*, 884–890. [CrossRef]

10. Kozlov, S.M.; Aleksandrov, H.A.; Neyman, K.M. Energetic Stability of Absorbed H in Pd and Pt Nanoparticles in a More Realistic Environment. *J. Phys. Chem. C* **2015**, *119*, 5180–5186. [CrossRef]

11. Crampton, A.S.; Rötzer, M.D.; Schweinberger, F.F.; Yoon, B.; Landman, U.; Heiz, U. Ethylene hydrogenation on supported Ni, Pd and Pt nanoparticles: Catalyst activity, deactivation and the d-band model. *J. Catal.* **2016**, *333*, 51–58. [CrossRef]

12. Zhang, J.; Mo, Y.; Vukmirovic, M.B.; Klie, R.; Sasaki, K.; Adzic, R.R. Platinum monolayer electrocatalysts for O_2 reduction: Pt monolayer on Pd(111) and on carbon-supported Pd nanoparticles. *J. Phys. Chem. B* **2004**, *108*, 10955–10964. [CrossRef]

13. Park, J.Y. *Current Trends of Surface Science and Catalysis*; Springer: New York, NY, USA, 2014.

14. Schulte, J. *Nanotechnology: Global Strategies, Industry Trends and Applications*; John Wiley & Sons Ltd.: Hoboken, NJ, USA, 2005.

15. Rakhshani, A.E.; Makdisi, Y.; Ramazaniyan, H.A. Electronic and optical properties of fluorine-doped tin oxide films. *J. Appl. Phys.* **1998**, *83*, 1049–1057. [CrossRef]

16. Elangovan, E.; Ramamurthi, K. A study on low cost-high conducting fluorine and antimony-doped tin oxide thin films. *Appl. Surf. Sci.* **2005**, *249*, 183–196. [CrossRef]

17. Ruffino, F.; Gentile, A.; Zimbone, M.; Piccitto, G.; Reitano, R.; Grimaldi, M.G. Size-selected Au nanoparticles on FTO substrate: Controlled synthesis by the Rayleigh-Taylor instability and optical properties. *Superlatt. Microstruct.* **2016**, *100*, 418–430. [CrossRef]

18. Ginley, D.S.; Hosono, H.; Paine, D.C. *Handbook of Transparent Conductors*; Springer: Berlin, Germany, 2010.

19. Kent, C.A.; Concepcion, J.J.; Dares, C.J.; Torelli, D.A.; Rieth, A.J.; Miller, A.S.; Meyer, T.J. Water oxidation and oxygen monitoring by cobalt-modified fluorine-doped TiN oxide electrodes. *J. Am. Chem. Soc.* **2013**, *135*, 8432–8435. [CrossRef] [PubMed]

20. Forte, K.; Serbin, J.; Koch, J.; Egbert, A.; Fallnich, C.; Ostendorf, A.; Chichkov, B.N. Towards nanostructuring with femtosecond laser pulse. *Appl. Phys. A* **2003**, *77*, 229–235.

21. Moening, J.P.; Thanawala, S.S.; Georgiev, D.G. Formation of high-aspect-ratio protrusions on gold films by localized pulsed laser irradiation. *Appl. Phys. A* **2009**, *95*, 635–638. [CrossRef]

22. Favazza, C.; Kalayanaraman, R.; Sureshkumar, R. Dynamics of ultrathin metal films on amorphous substrates under fast thermal processing. *J. Appl. Phys.* **2007**, *102*, 104308. [CrossRef]

23. Henley, S.J.; Carrey, J.D.; Silva, S.R.P. Metal nanoparticle production by pulsed laser nanostructuring of thin metal films. *Appl. Surf. Sci.* **2007**, *253*, 8080–8085. [CrossRef]

24. Trice, J.; Thomas, D.; Favazza, C.; Sureshkumar, R.; Kalyanaraman, R. Pulsed-laser induced dewetting in nanoscopic metal films: Theory and experiments. *Phys. Rev. B* **2007**, *75*, 235439. [CrossRef]

25. Ruffino, F.; Pugliara, A.; Carria, E.; Bongiorno, C.; Spinella, C.; Grimaldi, M.G. Formation of nanoparticles from laser irradiated Au thin films on SiO_2/Si: Elucidating the Rayleigh-instability role. *Mater. Lett.* **2012**, *84*, 27–30. [CrossRef]

26. Ruffino, F.; Carria, E.; Kimiagar, S.; Crupi, I.; Simone, F.; Grimaldi, M.G. Formation and evolution of nanoscale metal structures on ITO surface by nanosecond laser irradiations of thin Au and Ag films. *Sci. Adv. Mater.* **2012**, *4*, 708–718. [CrossRef]

27. Ruffino, F.; Grimaldi, M.G. Controlled dewetting as fabrication and patterning strategy for metal nanostructures. *Phys. Status Solidi A* **2015**, *212*, 1662–1684. [CrossRef]

28. González, A.G.; Diez, J.A.; Wu, Y.; Fowlkes, J.D.; Rack, P.D.; Kondic, L. Instability of liquid Cu films on a SiO_2 substrate. *Langmuir* **2013**, *29*, 9378–9387. [CrossRef]

29. Fowlkes, J.D.; Kondic, L.; Diez, J.; Rack, P.D. Self-assembly versus directed assembly of nanoparticles via pulsed laser induced dewetting of patterned metal films. *Nano Lett.* **2011**, *11*, 2478–2485. [CrossRef] [PubMed]

30. Ruffino, F.; Pugliara, A.; Carria, E.; Romano, L.; Bongiorno, C.; Fisicaro, G.; La Magna, A.; Spinella, C.; Grimaldi, M.G. Towards a laser fluence dependent nanostructuring of thin Au films on Si by nanosecond laser irradiation. *Appl. Surf. Sci.* **2012**, *258*, 9128–9137. [CrossRef]

31. Thompson, C.V. Solid-State Dewetting of Thin Films. *Annu. Rev. Mater. Res.* **2012**, *42*, 399–434. [CrossRef]

32. Kwon, J.-Y.; Yoon, T.-S.; Kim, K.-B.; Min, S.-H. Comparison of the agglomeration behavior of Au and Cu films sputter deposited on silicon dioxide. *J. Appl. Phys.* **2003**, *93*, 3270–3278. [CrossRef]

33. Giermann, A.L.; Thompson, C.V. Solid-state dewetting for ordered arrays of crystallographically oriented metal particles. *Appl. Phys. Lett.* **2005**, *86*, 121903. [CrossRef]

34. Wang, D.; Ji, R.; Schaaf, P. Formation of precise 2D Au particle arrays via thermally induced dewetting on pre-patterned substrates. *Beilstein J. Nanotechnol.* **2011**, *2*, 318–326. [CrossRef]

35. Wang, D.; Schaaf, P. Solid-state dewetting for fabrication of metallic nanoparticles and influences of nanostructured substrates and dealloying. *Phys. Status Solidi A* **2013**, *210*, 1544–1551. [CrossRef]

36. Yang, S.; Xu, F.; Ostendorp, S.; Wilde, G.; Zhao, H.; Lei, Y. Template-confined dewetting process to surface nanopatterns: Fabrication, structural tunability, and structure-related properties. *Adv. Funct. Mater.* **2011**, *21*, 2446–2455. [CrossRef]

37. Ruffino, F.; Grimaldi, M.G. Self-organized patterned arrays of Au and Ag nanoparticles by thickness-dependent dewetting of template-confined films. *J. Mater. Sci.* **2014**, *49*, 5714–5729. [CrossRef]

38. Wang, J.T.; Shi, X.L.; Liu, W.W.; Zhong, X.H.; Wang, J.N.; Pyrah, L.; Sanderson, K.D.; Ramsey, P.M.; Hirata, M.; Tsuri, K. Influence of preferred orientation on the electrical conductivity of fluorine-doped TiN oxide films. *Sci. Rep.* **2014**, *4*, 3679. [CrossRef] [PubMed]

39. Wang, J.T.; Shi, X.L.; Zhong, X.H.; Wang, J.N.; Pyrah, L.; Sanderson, K.D.; Ramsey, P.M.; Hirata, M.; Tsuri, K. Morphology control of fluorine-doped tin oxide thin films for enhanced light trapping. *Sol. Energy Mater. Sol. Cells* **2015**, *132*, 578–588. [CrossRef]

40. Ruffino, F. Experimental analysis on the molten-phase dewetting characteristics of AuPd Alloy films on topoghraphically-structured substrates. *Metals* **2017**, *7*, 327. [CrossRef]

41. Gentile, A.; Cacciato, G.; Ruffino, F.; Reitano, R.; Scapellato, G.; Zimbone, M.; Lombardo, M.; Battaglia, A.; Gerardi, C.; Foti, M.; et al. Nanoscale structuration and optical properties of thin gold films on textured FTO. *J. Mater. Sci.* **2014**, *49*, 8498–8507. [CrossRef]

42. Zhou, Z.; Song, Z.; Li, L.; Zhang, J.; Wang, Z. Fabrication of periodic variable-sized Pt nanoparticles via laser interference patterning. *Appl. Surf. Sci.* **2015**, *335*, 65–70. [CrossRef]

43. Owusu-Ansah, E.; Horwood, C.A.; El-Sayed, H.A.; Birss, V.I.; Shi, Y.J. A method for the formation of Pt metal nanoparticles using nanosecond pulsed laser dewetting. *Appl. Phys. Lett.* **2015**, *106*, 203103. [CrossRef]

44. Available online: http://www.kintec.hk/ (accessed on 10 November 2018).

45. Venables, J.A.; Spiller, G.D.; Hanbücken, M. Nucleation and growth of thin films. *Rep. Prog. Phys.* **1984**, *47*, 399–459. [CrossRef]

46. Ruffino, F.; Grimaldi, M.G. Island-to-percolation transition during the room-temperature growth of sputtered nanoscale Pd films on hexagonal SiC. *J. Appl. Phys.* **2010**, *107*, 074301. [CrossRef]

47. Zhang, L.; Cosandey, F.; Persaud, R.; Madey, T.E. Initial growth and morphology of thin Au films on $TiO_2(110)$. *Surf. Sci.* **1999**, *439*, 73–85. [CrossRef]

48. Li, Y.; Tang, C.; Zhong, J.; Meng, L. Dewetting and detachment of Pt nanofilms on graphitic substrates: A molecular dynamics study. *J. Appl. Phys.* **2015**, *117*, 064304. [CrossRef]

49. Font, F.; Afkhami, S.; Kondic, L. Substrate melting during laser heating of nanoscale metal films. *Int. J. Heat Mass Transf.* **2017**, *113*, 237–245. [CrossRef]

50. Poulier, C.; Smith, D.S.; Absi, J. Thermal conductivity of pressed powder compacts: Tin oxide and alumina. *J. Eur. Ceram. Soc.* **2007**, *27*, 475–478. [CrossRef]

51. Lu, L.-X.; Wang, Y.-M.; Srinivasan, B.M.; Asbahi, M.; Yang, J.K.W.; Zhang, Y.-W. Nanostructure formation by controlled dewetting on patterned substrates: A combined theoretical, modeling and experimental study. *Sci. Rep.* **2016**, *6*, 32398. [CrossRef] [PubMed]

52. Mukherjee, R.; Bandyopadhyay, D.; Sharma, A. Control of morphology in pattern directed dewetting of thin polymer films. *Soft Matter* **2008**, *4*, 2086–2097. [CrossRef]

53. Kargupta, K.; Sharma, A. Dewetting of thin films on periodic physically and chemically patterned surfaces. *Langmuir* **2002**, *18*, 1893–1903. [CrossRef]

54. Kargupta, K.; Sharma, A. Templating of thin films induced by dewetting on patterned surfaces. *Phys. Rev. Lett.* **2001**, *86*, 4536–4539. [CrossRef] [PubMed]

55. Liu, H.; Wang, A.; Sun, Q.; Wang, T.; Zeng, H. Cu Nanoparticles/fluorine-doped TiN oxide (FTO) nanocomposites for photocatalytic H_2 evolution under visible light irradiation. *Catalysts* **2017**, *7*, 385. [CrossRef]

56. Lu, X.; Rycenga, M.; Skrabalak, S.E.; Wiley, B.; Xia, Y. Chemical synthesis of novel plasmonic nanoparticles. *Annu. Rev. Phys. Chem.* **2009**, *60*, 167–192. [CrossRef] [PubMed]

57. Sugawa, K.; Tahara, H.; Yamashita, A.; Otsuki, J.; Sagara, T.; Harumoto, T.; Yanagida, S. Refractive index susceptibility of the plasmonic Palladium nanoparticle: Potential as the third plasmonic sensing material. *ACS Nano* **2015**, *9*, 1895–1904. [CrossRef] [PubMed]

58. Maier, S.A. *Plasmonics-Fundamental and Applications*; Springer: New York, NY, USA, 2007.

59. Windt, D.L.; Cash, W.C.; Scott, M.; Arendt, P.; Newnam, B.; Fisher, R.F.; Swartlander, A.B. Optical constants for thin films of Ti, Zr, Nb, Mo, Ru, Rh, Pd, Ag, Hf, Ta, W, Re, Ir, Os, Pt, and Au from 24 Å to 1216 Å. *Appl. Opt.* **1988**, *27*, 246–278. [CrossRef] [PubMed]

60. Rakić, A.D.; Djurišić, A.B.; Elazar, J.M.; Majewski, M.L. Optical properties of metallic films for vertical-cavity optoelectronic devices. *Appl. Opt.* **1998**, *37*, 5271–5283. [CrossRef] [PubMed]

61. Werner, W.S.M.; Glantschnig, K.; Ambrosch-Draxl, C. Optical constants and inelastic electron-scattering data for 17elemental metals. *J. Phys. Chem. Ref. Data* **2009**, *38*, 1013–1092. [CrossRef]

62. Turrell, G.; Corset, J. *Raman Microscopy: Developments and Applications*; Elsevier Academic Press: Amsterdam, The Netherlands, 1996.

63. Han, Y.; Lupitskyy, R.; Chou, T.-M.; Stafford, C.M.; Du, H.; Sukhishvili, S. Effect of Oxidation on Surface-Enhanced Raman Scattering Activity of Silver Nanoparticles: A Quantitative Correlation. *Anal. Chem.* **2011**, *83*, 5873–5880. [CrossRef]

64. Westerström, R.; Messing, M.E.; Blomberg, S.; Hellman, A.; Grönbeck, H.; Gustafson, J.; Martin, N.M.; Balmes, O.; van Rijn, R.; Andersen, J.N.; et al. Oxidation and reduction of Pd(100) and aerosol-deposited Pd nanoparticles. *Phys. Rev. B* **2011**, *83*, 115440. [CrossRef]

65. Yoshida, H.; Omote, H.; Takeda, S. Oxidation and reduction processes of platinum nanoparticles observed at the atomic scale by environmental transmission electron microscopy. *Nanoscale* **2014**, *6*, 13113–13118. [PubMed]
66. Kracher, M.; Worsch, C.; Rüssel, C. Optical properties of palladium nanoparticles under exposure of hydrogen and inert gas prepared by dewetting synthesis of thin-sputtered layers. *J. Nanopart. Res.* **2013**, *15*, 1594. [CrossRef]
67. Yin, Y.; Li, Z.-Y.; Zhong, Z.; Gates, B.; Xia, Y.; Venkateswaran, S. Synthesis and characterization of stable aqueous dispersions of silver nanoparticles through the Tollens process. *J. Mater. Chem.* **2002**, *12*, 522–527. [CrossRef]
68. Lok, C.-N.; Ho, C.-M.; Chen, R.; He, Q.-Y.; Yu, W.-Y.; Sun, H.; Tam, P.K.-H.; Chiu, J.-F.; Che, C.-M. Silver nanoparticles: Partial oxidation and antibacterial activities. *J. Biol. Inorg. Chem.* **2007**, *12*, 527–534. [CrossRef]
69. Reguera, J.; Langer, J.; Jiménez de Aberasturi, D.; Liz-Marzán, L.M. Anisotropic metal nanoparticles for surface enhanced Raman scattering. *Chem. Soc. Rev.* **2017**, *46*, 3866–3885. [CrossRef] [PubMed]

© 2019 by the authors. Licensee MDPI, Basel, Switzerland. This article is an open access article distributed under the terms and conditions of the Creative Commons Attribution (CC BY) license (http://creativecommons.org/licenses/by/4.0/).

 *coatings*

Article

Synthesis and Deposition of Ag Nanoparticles by Combining Laser Ablation and Electrophoretic Deposition Techniques

Mònica Fernández-Arias [1], Massimo Zimbone [2,*], Mohamed Boutinguiza [1], Jesús Del Val [1], Antonio Riveiro [1], Vittorio Privitera [2], Maria G. Grimaldi [2,3] and Juan Pou [1]

[1]   Department of Applied Physics, University of Vigo, EEI, Lagoas-Marcosende, 36310 Vigo, Spain
[2]   Institute for Microelectronics and Microsystems, National Research Council (CNR-IMM), via S. Sofia 64, 95123 Catania, Italy
[3]   Dipartimento di Fisica e Astronomia, Università di Catania, via S. Sofia 64, 95123 Catania, Italy
*   Correspondence: massimo.zimbone@ct.infn.it; Tel.: +39-095-378-5350

Received: 30 July 2019; Accepted: 3 September 2019; Published: 6 September 2019

Abstract: Silver nanostructured thin films have been fabricated on silicon substrate by combining simultaneously pulsed laser ablation in liquid (PLAL) and electrophoretic deposition (ED) techniques. The composition, topography, crystalline structure, surface topography, and optical properties of the obtained films have been studied by energy dispersive X-ray spectroscopy (EDS), high-resolution transmission electron microscopy (HRTEM), X-ray diffraction (XRD), and UV-visible spectrophotometry. The coatings were composed of Ag nanoparticles ranging from a few to hundred nm. The films exhibited homogenous morphology, uniform appearance, and a clear localized surface plasmon resonance (LSPR) around 400 nm.

Keywords: silver nanoparticles; electrophoretic deposition; pulsed laser ablation in liquid

---

1. Introduction

Noble metal nanoparticles in solution and deposited as thin films have attained wide popularity in the last 10 years and aroused intense research interest in nanotechnology due to their well-known properties, such as good conductivity, localized surface plasmon resonance (LSPR), and antibacterial and catalytic effects [1–4]. They are used in many different areas, such as medicine, solar cells, nano- and microelectronics, scientific investigations [5]. Films based on noble metals are the object of intense investigation due to the optical properties introduced by the characteristic LSPR, which generates an optical local field enhancement. This collective coherent oscillation of electrons on the conductive band of metallic nanoparticles interacts with the electromagnetic field and produces a strong absorption in particular regions of the electromagnetic spectrum. In particular, silver nanoparticles can absorb light in the near UV region of the spectrum extending their wavelength response to visible light. Other metal nanoparticles than noble metals also show absorption in visible light, such as Cu and Al [6,7], but their LSPR is weak and suffers from relatively easy oxidation. The most intensely studied plasmonic nanoparticle materials are Au and Ag. Ag is cheaper than Au and presents an even stronger LSPR, making Ag nanoparticles good candidates for many different applications. They can be used for enhancing solar cell efficiency by means of light trapping in organic as well as inorganic solar cells [8,9] or they can be used together with TiO_2 to improve its photocatalytic activity. These nanoparticles produce an increased absorption in the visible region due to their aforementioned LSPR [10]. The very high electric field around the nanoparticles makes them very good candidates for enhancing the signal of Raman scattering spectroscopy (SERS) [11,12], luminescence, and cathodoluminescence [13,14].

A wide range of techniques for producing nanoparticles is used in the literature, and can be roughly classified into two groups based on chemical or physical processes. The most popular method is chemical reduction, whereby organic or inorganic agents are used to reduce Ag salt and generate nanoparticles with different sizes, shapes, and composition [15–19]. Most of these methods use chemical reactants, surfactants, or stabilizers which contaminate the final nanoparticles, necessitating further extraction and purification [17]. It is worth pointing out that this is one of the main issues with nanoparticle technology. Physical processes, on the contrary, use metallic precursors and allow nanomaterials to be obtained without the presence of chemical reagents, but size and shape control is challenging. Among the physical techniques, laser-based ones have been used extensively. In particular, pulsed laser ablation in liquid (PLAL) is one of the most promising techniques to produce very pure nanoparticles. Indeed, it avoids surfactants and reaction products in solution. In PLAL, a high fluence laser pulse impinges on a solid substrate immersed in liquid. Plasma plume is formed and expands into liquid. The rapid cooling realizes nanoparticles in liquids. In previous works, we have obtained ceramic and metal nanoparticles in water and in open air [20–24]. Moreover, nanoparticles obtained by PLAL are charged and, once synthesized in solution, can be electrodeposited in a suitable substrate. There are different methods for nanoparticle deposition, such as chemical vapor deposition (CVD), physical vapor deposition (PVD), pulsed laser deposition (PLD), and sputtering. While most of these require sophisticated and expensive equipment [23–25], electrophoretic deposition (ED) is a simple and low-cost technique for nanomaterial deposition. This method has already been used to deposit Ag nanoparticles prepared by chemical methods [26]. It allows the synthesis parameters of nanoparticles to be controlled and incorporated into materials, which enables materials to be obtained and fabricated with tailored properties.

In this work, we report the results of combining the PLAL with ED technologies to obtain films of Ag nanoparticles deposited on a substrate. By combining a synthesis technique and a deposition method, Ag nanoparticles are produced and deposited in one step. This process will have great advantages in terms of costs and time, allowing a spread use of the combined technique.

2. Materials and Methods

Plates of Ag of 99.99% purity were cleaned and sonicated for ablation by laser in water. The targets were attached to the bottom of a glass box and filled with distilled water up to 1.0 mm over the upper surface of the Ag plate. The Ag targets were used as positive electrodes while a substrate of Si separated 2.5 cm was used as negative electrodes, as shown in Figure 1.

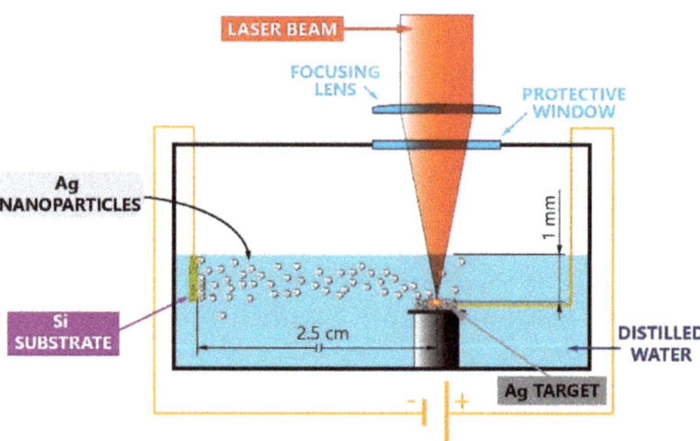

Figure 1. Schematic illustration of the experimental setup used to synthesize and deposit Ag nanoparticles on Si. Note: dimensions not to scale.

The laser source used consisted of a pulsed diode-pumped Nd:YVO4 laser (Rofin, Hamburg, Germany). It provides laser pulses at 532 nm with pulse duration of 14 ns, a repetition rate of 20 kHz, and an average output power of 6.0 W. The Ag was ablated in water while an electric field of 15 V was applied between the electrodes.

Scanning electron microscopy (SEM) was used to observe and analyze the surface morphology and the microstructure by means of a JSM-6700 field emission scanning electron microscope (JEOL, Tokyo, Japan). Transmission electron microscopy (TEM), selected area electron diffraction (SAED), and high-resolution transmission electron microscopy (HRTEM) images were taken on a JEOL JEM-2010 FEG transmission electron microscope equipped with a slow digital camera scan, using an accelerating voltage of 200 kV, and provided by an energy dispersive X-ray spectrometer (EDS) to reveal their crystallinity and composition. The optical absorption of Ag nanoparticles deposited on the glass was measured by UV-Vis in a Hewlett Packard HP 8452 spectrophotometer (San Jose, CA, USA) in the wavelength range of 190–800 nm in a 10 mm quartz cell. Electron energy loss spectroscopy (EELS, JEOL, Tokyo, Japan) was carried out in order to check the potential presence of oxygen in the synthesized and deposited nanoparticles.

3. Results and Discussion

When a high-power nanosecond laser beam strikes on the Ag target, the local temperature rises above the Ag boiling point, leading to the formation of a plasma plume with the presence of different species such as atoms and ions. These species are confined by the surrounding liquid and remain inside the cavitation bubble realized by the evaporated solvent. Nanoparticles nucleate and grow by coalescence as the plasma cools down [27]. The obtained nanoparticles are charged and, due to the presence of an electric field inside the solution, can be accelerated and deposited on a suitable electrode, forming a thin film.

We analyzed both nanoparticles in solution and deposited on the substrate. Figure 2 exhibits the typical TEM image of the nanoparticles found in solution.

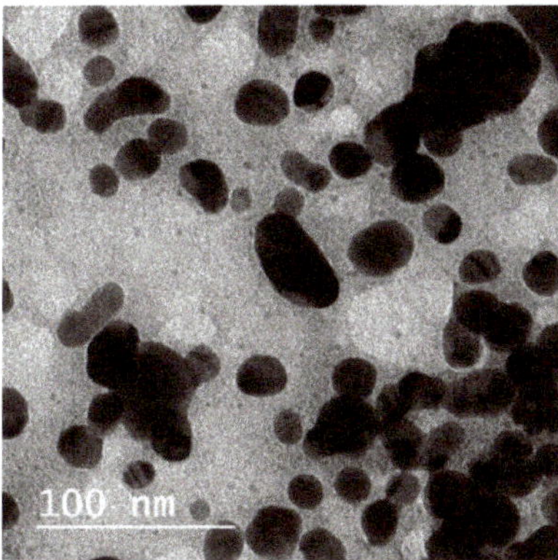

Figure 2. Transmission electron microscopy (TEM) micrograph of the obtained nanoparticles with an elongated shape.

The nanoparticles are ellipsoidal in shape. It is worth pointing out that they are not spherical as obtained in previous works [28,29]. The ellipsoidal shape of the nanoparticles is probably favored by the presence of the electric field, which contributes to aligning nanoparticles during their formation. Nanoparticles show broad distribution in terms of both size and shape. The agglomeration is probably produced during the deposition of the solution drop on the TEM grid.

Figure 3 shows the selected area electron diffraction (SAED) performed on different groups of particles. It shows the concentric diffraction rings with bright spots.

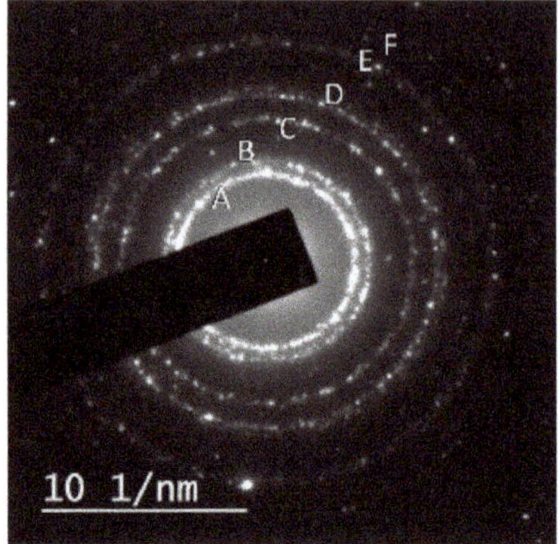

Figure 3. Selected area electron diffraction (SAED) pattern exhibiting diffraction rings obtained from Ag nanoparticles. The diffraction peaks from A, B, C, D, E, and F are consistent with Ag, showing 0.235, 0.204, 0.144, 0.123, 0.118, 0.101 nm and the corresponding {111}, {200}, {220}, {311}, {222}, {400} family planes of silver.

We observed that all the obtained particles are crystalline. The measured interplanar distances are listed in Table 1 and compared to those of metallic silver and Ag_2O.

Table 1. The d-spacing as measured from selected area electron diffraction (SAED) performed on Ag nanoparticles obtained by laser ablation of Ag target in water using 532 nm laser compared to those of metallic Ag and Ag_2O.

d_{hkl} (nm)	A	B	C	D	E	F
Measured	0.235	0.204	0.144	0.123	0.118	0.101
Ag JCPDS_ICDD (1993)	0.236	0.204	0.145	0.123	0.118	0.102
Ag_2O JCPDS_ICDD (1993)	0.237	–	0.143	–	0.118	–

The planar distances measured from the synthesized nanoparticles show good agreement with metallic silver even though the nucleation of the nanoparticle is performed in the presence of evaporated water. Their interplanar distances were calculated to be 0.235, 0.204, 0.144, 0.123, 0.118 and 0.101 nm, which could be assigned to the distances between {111}, {200}, {220}, {311}, {222} and {400} family planes of cubic silver.

To elucidate the crystalline nature of the nanoparticles more precisely, further investigations were made on single particles by means of HRTEM, as can be observed from Figure 4a, where clear lattice fringes with the d-spacing values are displayed.

The d-spacing measured from the corresponding fast Fourier transform (FFT) (Figure 4b) gives 0.205 and 0.238 nm which could be assigned to the {200} and {111} family planes of cubic silver whose corresponding interplanar distances are 0.204 and 0.236 nm respectively, while 0.238 nm could also be attributed to the {200} family planes of Ag_2O. The high magnification image in Figure 4a of the produced nanoparticle does not reveal any crystalline discontinuity or amorphous nature on the shell and no apparent oxide layer was found.

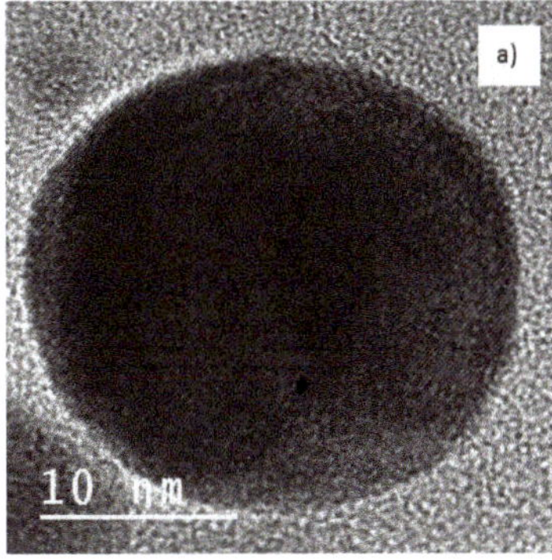
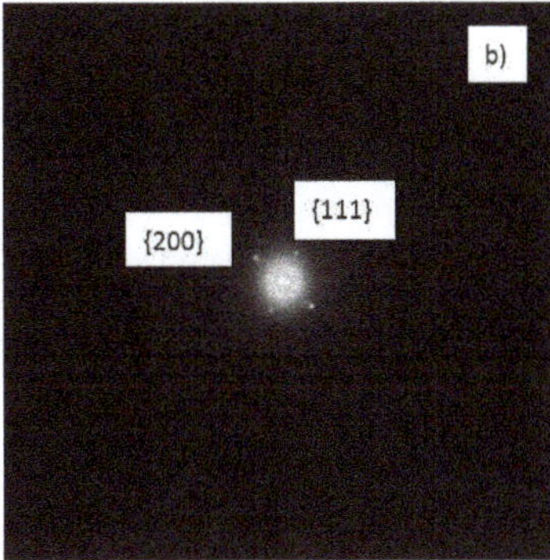

Figure 4. High-resolution transmission electron microscopy (HRTEM) of single particles showing clear fringes (**a**) with the corresponding fast Fourier transform (FFT) (**b**) whose d-spacing, 0.205 and 0.238 nm, could be assigned to the {200} and {111} family planes of cubic silver.

Together with the important specific area of the nanoparticles, their LSPR show dependence on their size and shape as well as the dielectric properties of the surrounding media [30,31]. The nanoparticles exhibit irregular shape rather than spherical morphology, which can favor the excitation of higher-order plasmon modes over small nanospheres. When the surface plasmon is excited, the intensity of the electromagnetic field around the sharp edges of nanoparticles can be significantly greater than the incident radiation [32]. Figure 5 shows the measured absorption spectrum of silver nanoparticles by laser irradiation of Ag submerged in water.

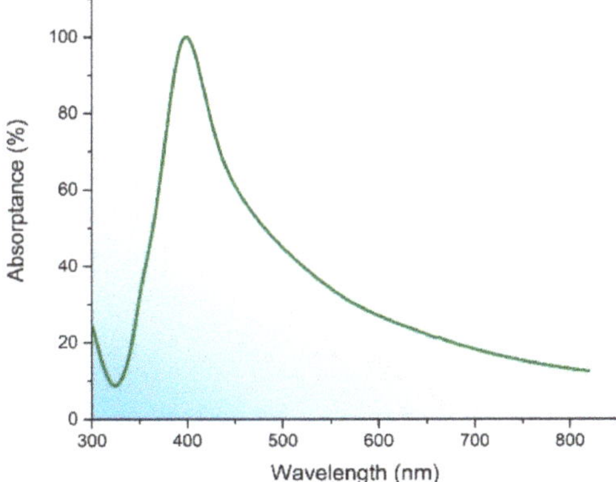

Figure 5. UV-Vis absorption spectrum of silver nanoparticles obtained by laser ablation of Ag target submerged in water with LSPR around 400 nm.

The UV-Vis absorption spectrum of the synthesized silver nanoparticles and grown-over Si substrate in the present work exhibits a plasmon resonance band around 400 nm, similar to characteristic LSPR of nanometric spherical particles [33]. The peak around 400 nm corresponds to dipolar mode excitation of spherical silver nanoparticles, but there is no significant position shifting of LSPR due to their size increasing from a few nanometers to tens of nanometers [34]. The large full width at half maximum (FWHM) is attributable to large polydispersity which promotes wider absorbance bands [35]. Although the nanoparticles synthesized by laser ablation in the solution can form stable colloidal solution, the nanoparticles can form networks or small assemblies [36] due to the attractive van der Waal forces in the absence of capping agents.

We now analyze the nanoparticles deposited on the Si substrate. For low electric fields (below 2.0 V/cm), no significant amount of particles collected on the cathode. When the voltage was increased to 15 V (approx. 6 V/cm), however, nanoparticle deposition on the cathode is observed. Figures 6 and 7 show SEM images of the deposited nanoparticles. It is worth noting that the final coating density, aspect, and shape are highly affected by the size and the ablated nanoparticles. To obtain a uniform coating on Si substrate it is important to limit the nanoparticles' size distribution, which is mainly related to the pulse duration and the laser power. Decreasing the pulse width or keeping the laser power (pulse energy) slightly above the ablation threshold leads to more uniform coatings. By increasing the deposition time, the coating density is increased but the uniformity is decreased.

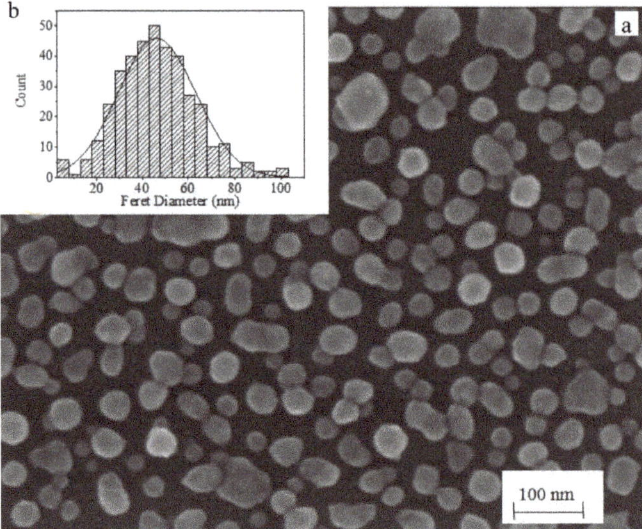

Figure 6. SEM micrograph (**a**) showing the distribution of the nanoparticles deposited on Si substrate with no agglomeration and the presence of slits among them. In (**b**), the Feret diameter distribution is shown.

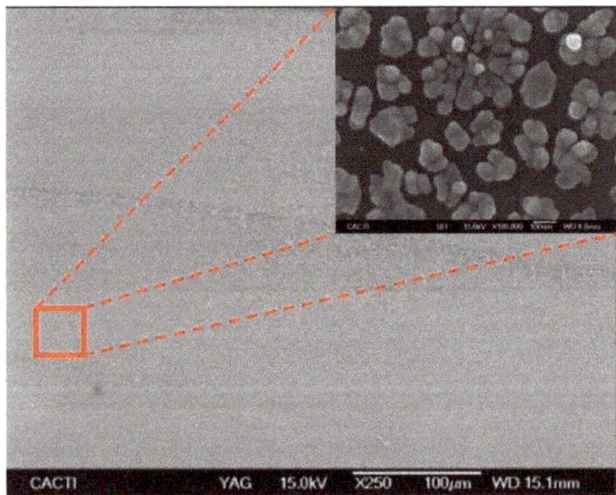

Figure 7. SEM micrograph showing the uniform film of the nanoparticles deposited on the Si substrate. The magnification in the inset illustrates the block formation on Si by the electric field.

We show the Feret diameter distribution in the inset of Figure 6. It is apparent that the size distribution on the cathode surface is uniform and monomodal. Monomodal distribution of nanoparticles can be explained by the Stokes law equation [37], which describes the mobility of small spherical nanoparticles in laminar flow under the action of electrical deriving force: $-v = qE/6\pi R\eta$. where q is the particle charge, E is the electrical field, R is the nanoparticle hydrodynamic radius, η is the water viscosity, and v the nanoparticle velocity. Despite energy-dispersive size distribution and non-spherical shape of the nanoparticles, their R/q ratio is very similar due to their reduced size, so the mobility and the distribution of nanoparticles on the substrate are mainly governed by the applied voltage, leading to a similar horizontal component velocity for particles. Gravity mainly affects bigger particles, deflecting their trajectories downward to not reach the cathode. On the other hand, the presence of an electrical field accelerates the charged nanoparticles to the cathode, contributing to the collision of particles of different sizes and the formation of assemblies. Most probably, the oversized blocks "sinking" before reaching the cathode.

The characteristics of the particles forming the obtained films are very important in terms of specific surface area and size, as well as their distribution on the substrate in order to enhance the efficiency of the incoming radiation, especially for peculiar applications such as photocatalysis or SERS. Indeed, the intensity of the electromagnetic field can be enhanced locally due to the interaction of plasmonic nanoparticles. Interaction enhances the electric field when the distance between nanoparticles is lower than the size of the particles. Indeed, strong electromagnetic enhancement can be detected in the gaps between nanoparticles when they are close to each other [38]. The deposited nanoparticles on Si, in this work, adopt the configuration mentioned above, showing slits among them smaller than the nanoparticle size, as can be seen from the SEM image in Figure 6. This fact makes the synthetized substrate suitable for SERS and photocatalytical applications.

4. Conclusions

In summary, we have synthesized silver nanoparticles and deposited them on Si substrate in a one-step process by combining the techniques of laser ablation of solids in liquids and ED to obtain uniform films. The silver target submerged in water was connected to the positive electrode and separated 2.0 cm from the cathode. The resulting nanostructured silver films consisted of homogenous coatings composed of irregular nanoparticles with slight slits among them. These films can be used to enhance the efficiency of photovoltaic devices, photocatalysis, or SERS.

Author Contributions: Conceptualization, Methodology, Validation and Writing—Original Draft Preparation: M.F.-A., M.Z., M.B., J.D.V. and A.R.; Writing—Review and Editing: V.P. and M.G.G.; Writing—Review, Editing and Supervision: J.P.

Funding: This work was partially supported by the EU research project Bluehuman (EAPA_151/2016 Interreg Atlantic Area), Government of Spain [RTI2018-095490-J-I00 (MCIU/AEI/FEDER, UE), Mobility Grant of Senior Professors and Researchers (Grant PRX15/00088)], and by Xunta de Galicia (ED431C 2019/23, ED481D 2017/010, ED481B 2016/047-0).

Acknowledgments: The technical staff from CACTI (University of Vigo) is gratefully acknowledged.

Conflicts of Interest: The authors declare no conflict of interest. The funders had no role in the design of the study; in the collection, analyses, or interpretation of data; in the writing of the manuscript, or in the decision to publish the results.

References

1. Krutyakov, Y.A.; Kudrinskiy, A.A.; Olenin, A.Y.; Lisichkin, G.V. Synthesis and properties of silver nanoparticles: Advances and prospects. *Russ. Chem. Rev.* **2008**, *77*, 233–257. [CrossRef]
2. Bok, Y.A.; Duoss, E.B.; Motala, M.J.; Guo, X.; Park, S.I.; Xiong, Y.; Yoon, J.; Nuzzo, R.G.; Rogers, J.A.; Lewis, J.A. Omnidirectional printing of flexible, stretchable, and spanning silver microelectrodes. *Science* **2009**, *323*, 1590–1593.
3. Shen, W.; Zhang, X.; Huang, Q.; Xu, Q.; Song, W. Preparation of solid silver nanoparticles for inkjet printed flexible electronics with high conductivity. *Nanoscale* **2014**, *6*, 1622–1628. [CrossRef] [PubMed]

4.   Raza, M.A.; Kanwal, Z.; Rauf, A.; Sabri, A.N.; Raiz, S.; Naseem, S. Size- and shape-dependent antibacterial studies of silver nanoparticles synthesized by wet chemical routes. *Nanomaterials* **2016**, *6*, 74. [CrossRef] [PubMed]

5.   Tran, Q.H.; Nguyen, V.Q.; Le, A.T. Silver nanoparticles: Synthesis, properties, toxicology, applications and perspectives. *Adv. Nat. Sci. Nanosci. Nanotechnol.* **2013**, *4*, 033001. [CrossRef]

6.   Buonsanti, R.; Llordes, A.; Aloni, S.; Helms, B.A.; Milliron, D.J. Tunable infraredabsorption and visible transparency of colloidal aluminum-doped zinc oxidenanocrystals. *Nano Lett.* **2011**, *11*, 4706–4710. [CrossRef]

7.   Serna, R.; Suarez-Garcia, A.; Afonso, C.N.; Babonneau, R. Optical evidence for reactive processes when embedding Cu nanoparticles in Al_2O_3 by pulsed laserdeposition. *Nanotechnology* **2006**, *17*, 4588–4593. [CrossRef]

8.   Liu, X.H.; Hou, L.X.; Wang, J.F.; Liu, B.; Yu, Z.S.; Ma, L.Q.; Yang, S.P.; Fu, G.S. Plasmonic-enhanced polymer solar cells with high efficiency by addition of silver nanoparticles of different sizes in different layers. *Sol. Energy* **2014**, *110*, 627–635. [CrossRef]

9.   Ho, W.J.; Lee, Y.Y.; Su, S.Y. External quantum efficiency response of thin silicon solar cell based on plasmonic scattering of indium and silver nanoparticles. *Nanoscale Res. Lett.* **2014**, *9*, 483. [CrossRef]

10.   Albitera, E.; Valenzuela, M.A.; Alfaro, S.; Valverde-Aguilar, G.; Martínez-Pallares, F.M. Photocatalytic deposition of Ag nanoparticles on TiO_2: Metal precursor effect on the structural and photoactivity properties. *J. Saudi Chem. Soc.* **2015**, *19*, 563–573. [CrossRef]

11.   Liu, F.X.; Tang, C.J.; Zhan, P.; Chen, Z.; Ma, H.T.; Wang, Z.L. Released plasmonic electric field of ultrathin tetrahedral-amorphous-carbon films coated Ag nanoparticles for SERS. *Sci. Rep.* **2013**, *4*, 4494. [CrossRef] [PubMed]

12.   Mondal, B.; Saha, S.K. Fabrication of SERS substrate using nanoporous anodic alumina template decorated by silver nanoparticles. *Chem. Phys. Lett.* **2010**, *497*, 89–93. [CrossRef]

13.   Lee, S.M.; Choi, K.C.; Kim, D.H.; Jeon, D.Y. Localized surface plasmon enhanced cathodoluminescence from Eu^{3+}-doped phosphor near the nanoscaled silverparticles. *Opt. Express* **2010**, *19*, 13209–13217. [CrossRef] [PubMed]

14.   Zhang, A.; Zhang, J.; Fang, Y. Photoluminescence from colloidal silver nanoparticles. *J. Lumin.* **2008**, *128*, 1635–1640. [CrossRef]

15.   Wang, H.; Qiao, X.; Chen, J.; Ding, S. Preparation of silver nanoparticles by chemical reduction method. *Colloids Surf. A: Physicochem. Eng. Asp.* **2005**, *256*, 111–115. [CrossRef]

16.   Hussain, J.I.; Kumar, S.; Hashmi, A.A.; Khan, Z. Silver nanoparticles: Preparation, characterization, and kinetics. *Adv. Mater. Lett.* **2011**, *2*, 188–194. [CrossRef]

17.   Guzmán, M.G.; Dille, J.; Godet, S. Synthesis of silver nanoparticles by chemical reduction method and their antibacterial activity. *IJCBE* **2009**, *2*, 104–111.

18.   Athanasiou, C.E.; Bellouard, Y. A monolithic micro-tensile tester for investigating silicon dioxide polymorph micromechanics, fabricated and operated using a femtosecond laser. *Micromachines* **2015**, *6*, 1365–1386. [CrossRef]

19.   Zhang, J.; Gecevičius, M.; Beresna, M.; Kazansky, P.G. Seemingly unlimited lifetime data storage in nanostructured glass. *Phys. Rev. Lett.* **2014**, *112*, 033901. [CrossRef]

20.   Boutinguiza, M.; Lusquiños, F.; Riveiro, A.; Comesaña, R.; Pou, J. Hydroxylapatite nanoparticles obtained by fiber laser induced fracture. *Appl. Surf. Sci.* **2009**, *255*, 5382–5385. [CrossRef]

21.   Boutinguiza, M.; Rodríguez-González, B.; del Val, J.; Comesaña, R.; Lusquiños, F.; Pou, J. Laser-assisted production of spherical TiO_2 nanoparticles in water. *Nanotechnology* **2011**, *22*, 195606. [CrossRef] [PubMed]

22.   Boutinguiza, M.; Comesaña, R.; Lusquiños, F.; Riveiro, A.; del Val, J.; Pou, J. Palladium nanoparticles produced by CW and pulsed laser ablation in water. *Appl. Surf. Sci.* **2014**, *302*, 19–23. [CrossRef]

23.   Okada, K. Plasma-enhanced chemical vapor deposition of nanocrystalline diamond. *Sci. Technol. Adv. Mater.* **2007**, *8*, 624–634. [CrossRef]

24.   Kong, Y.C.; Yu, D.P.; Zhang, B.; Fang, W.; Feng, S.Q. Ultraviolet-emitting ZnO nanowires synthesized by a physical vapor deposition approach. *Appl. Phys. Lett.* **2011**, *78*, 407–409. [CrossRef]

25.   Brodsky, M.H.; Cardona, M.; Cuomo, J.J. Infrared and Raman spectra of the silicon-hydrogen bonds in amorphous silicon prepared by glow discharge and sputtering. *Phys. Rev. B* **1977**, *16*, 3556–3571. [CrossRef]

26. López, I.; Vázquez, A.; Hernández-Padrón, G.H.; Gómez, I. Electrophoretic deposition (EPD) of silver nanoparticles and their application as surface-enhanced Raman scattering (SERS) substrates. *Appl. Surf. Sci.* **2013**, *280*, 715–719. [CrossRef]

27. Yang, G.W. Laser ablation in liquids: Applications in the synthesis of nanocrystals. *Prog. Mater. Sci.* **2007**, *52*, 648–698. [CrossRef]

28. Boutinguiza, M.; Comesaña, R.; Lusquiños, F.; Riveiro, A.; del Val, J.; Pou, J. Production of silver nanoparticles by laser ablation in open air. *Appl. Surf. Sci.* **2015**, *336*, 108–111. [CrossRef]

29. Fernández-Arias, M.; Boutinguiza, M.; del Val, J.; Medina, E.; Rodríguez, D.; Riveiro, A.; Comesaña, R.; Lusquiños, F.; Gil, F.J.; Pou, J. Re-irradiation of silver nanoparticles obtained by laser ablation in water and assessment of their antibacterial effect. *Appl. Surf. Sci.* **2019**, *473*, 548–554. [CrossRef]

30. Mejac, I.; Bryan, W.W.; Lee, T.R.; Tran, C.D. Visualizing the size, shape, morphology, and localized surface plasmon resonance of individual gold nanoshells by near-infrared multispectral imaging microscopy. *Anal. Chem.* **2009**, *81*, 6687–6694. [CrossRef]

31. Petryayeva, E.; Krull, U.J. Localized surface plasmon resonance: Nanostructures, bioassays and biosensing—A review. *Anal. Chim. Acta.* **2011**, *706*, 8–24. [CrossRef] [PubMed]

32. Hao, E.; Schatz, G.C.; Hupp, J.T. Synthesis and optical properties of anisotropic metal nanoparticles. *J. Fluoresc.* **2004**, *14*, 331–341. [CrossRef] [PubMed]

33. Tsuji, T.; Thanga, D.H.; Okazaki, Y.; Nakanishi, M.; Tsuboi, Y.; Tsuji, M. Preparation of silver nanoparticles by laser ablation in polyvinylpyrrolidone solutions. *Appl. Surf. Sci.* **2008**, *254*, 5224–5230. [CrossRef]

34. Krajczewski, J.; Kołataj, K.; Kudelski, A. Plasmonic nanoparticles in chemical analysis. *RSC Adv.* **2017**, *7*, 17559–17576. [CrossRef]

35. Oliveira, J.P.; Prado, A.P.; Juvêncio Keijok, W.; Ribeiro, M.R.N.; Pontes, M.J.; Nogueira, B.V.; Guimarães, M.C. A helpful method for controlled synthesis of monodisperse gold nanoparticles through response surface modelling. *Arab. J. Chem.* **2017**. [CrossRef]

36. Rong, W.; Ding, W.; Madler, L.; Ruoff, R.S.; Friedlander, S.K. Mechanical properties of nanoparticle chain aggregates by combined AFM and SEM: Isolated aggregates and networks. *Nano Lett.* **2006**, *6*, 2646–2655. [CrossRef] [PubMed]

37. Hiemenz, P.C.; Rajagopalan, R. *Principles of Colloid and Surface Chemistry*, 3rd ed.; Marcel Dekker, Inc.: New York, NY, USA, 1997; ISBN 9780824793975.

38. Yang, J.; Ren, F.; Chong, X.; Fan, D.; Chakravarty, S.; Wang, Z.; Chen, R.T.; Wang, A.X. Photonics guided-mode resonance grating with self-assembled silver nanoparticles for surface-enhanced Raman scattering spectroscopy. *Photonics* **2014**, *1*, 380–389. [CrossRef] [PubMed]

© 2019 by the authors. Licensee MDPI, Basel, Switzerland. This article is an open access article distributed under the terms and conditions of the Creative Commons Attribution (CC BY) license (http://creativecommons.org/licenses/by/4.0/).

Article

Laser Surface Texturing of Alumina/Zirconia Composite Ceramics for Potential Use in Hip Joint Prosthesis

Francesco Baino [1],*, Maria Angeles Montealegre [2], Joaquim Minguella-Canela [3] and Chiara Vitale-Brovarone [1]

[1] Department of Applied Science and Technology, Politecnico di Torino, Corso Duca degli Abruzzi 24, 10129 Torino, Italy; chiara.vitale@polito.it
[2] AIMEN Technology Centre, Relva 27A Torneiros, 36410 Porriño, Spain; mmontealegre@outlook.es
[3] Centre CIM, Departament d'Enginyeria Mecànica, Universitat Politècnica de Catalunya, Av. Diagonal, 647, 08028 Barcelona, Spain; joaquim.minguella@upc.edu
* Correspondence: francesco.baino@polito.it; Tel.: +39-011-090-4668

Received: 18 April 2019; Accepted: 5 June 2019; Published: 6 June 2019

Abstract: The use of metal shell to fix an acetabular cup to bone in hip joint prosthesis carries some limitations, including restrictions in prosthetic femur ball diameter and in patient's range of motion. These drawbacks could be ideally overcome by using a monolithic ceramic acetabular cup, but the fixation of such an implant to host bone still remains a challenge. Since porous surfaces are known to promote more bone tissue interlocking compared to smooth materials, in this work the surfaces of sintered alumina/zirconia composite ceramics were treated by a pulsed laser radiation at 1064 nm with a pulse width in the nanosecond range, in order to impart controlled textural patterns. The influence of laser process parameters (e.g., energy per pulse, repetition rate, scanning speed, repetition number, angle of laser beam, and number of cycles) on the roughness and texture orientation was systematically investigated. The obtained surface topographies were inspected by optical and scanning electron microscopy, and the roughness was assessed by contact profilometry. Surface roughness could be modulated in the range of 3 to 30 μm by varying the processing parameters, among which the number of cycles was shown to play a major role. The laser treatment was also successfully adapted and applied to ceramic acetabular cups with a curved profile, thus demonstrating the feasibility of the proposed approach to process real prosthetic components.

Keywords: bioceramics; laser ablation; roughness; composites; hip joint prosthesis; cementless cup; bone

---

1. Introduction

The hip joint prosthesis has been the most active field of research in prosthetics over the last decades [1]. This high interest was mainly dictated by the increasing demand for such an implant in the elderly due the increment of the aging population worldwide [2]. Both the acetabular component and the femur stem are often fixed to the patient's bone using acrylic cement; however, this approach, despite the ease of execution, suffers from some drawbacks. Firstly, the in-situ cement polymerization reaction is typically associated to a sudden increase of local temperature that can induce bone tissue necrosis. Secondly, the cement is prone to disintegrate over time and produces debris, which may yield implant mobilization and bone necrosis at the operation site in the long term [3]. Therefore, several prosthetic models have been developed in the attempt to overcome the need for acrylic cement while ensuring satisfactory and safe performance of the implant [4–6]. This is particularly challenging if adopting ceramic-on-ceramic bearings is a goal, given that such combinations are the most favorable in regard to excellent bio-inertness and minimal wear [7].

Cementless ceramic acetabular components were introduced for the first time in total hip arthroplasty in the 1970s, and monolithic ceramic screwed cups were typically combined with cemented femur stems having a ceramic ball head [8]. In 1999, Griss et al. [9] presented the long-term results of alumina screwed cylindrical implants and reported that 24 cases out of 67 underwent surgical revision due to loosening and migration of the acetabular cup. Another 25 cups were still in situ after 22 years of implantation, but all of them were found to have migrated to a certain extent under radiographic investigation.

Gierse et al. [10] also reported an initial or short-term postoperative migration in 35% of the patients receiving this ceramic cup.

Monolithic ceramic screwed cups with a conical shape were also proposed (the so-called Mittelmeier design), but high rates of implant loosening (53% after 12.3 years of postoperative follow-up) were reported [11].

The clinical use of all these screwed ceramic cups was soon stopped due to the above-mentioned postoperative problems and, at present, ceramic-on-ceramic bearing is based on a "modular" concept, i.e., a metal shell is used as a press-fit cementless acetabular component combined with a ceramic insert [12].

In principle, this design style suffers from a major disadvantage compared to monolithic cups, i.e., the inherent minimum thickness of metal shell and insert together, which restricts the use of both small-sized cups and large-diameter femur heads. Furthermore, problems of incorrect positioning (miscentering) of the ceramic insert into the metal shell might occur during surgery.

After a hiatus of about 20 years, the idea of fabricating monolithic ceramic cups able to be somehow anchored to the pelvic bone without using acrylic cement or metal shell was resurrected by Schreiner et al. [13], who pointed out the beneficial effect of surface porosity for allowing osteoconduction. This property was first reported by Forgon et al. [14] who observed bone ingrowth into 500 μm holes drilled in alumina implants. As discussed later by Cornell and Lane [15], osteoconduction can be observed in biomaterials that have a porosity similar to that of bone architecture. In fact, although growth factors, local cytokines, and other biomolecules modulate the process—which also strongly depends on the composition of the implanted material—the three-dimensional (3D) structure of an implant still plays a key role in affecting the host biological responses.

In their study published in 2011, Schreiner et al. [13] evaluated the osteointegration of alumina/zirconia composite cups (BIOLOX delta®, CeramTec, Plochingen, Germany) with a porous outer surface after implantation in a sheep model for 8 and 52 weeks. Stable osteointegration was detected at both time points and the interlocking of bone tissue into the implant pores was claimed to provide a stable fixation of the cup in the presence of low osteointegration rates. Since then, investigations on this material were apparently discontinued; this was probably due to commercial reasons, as other studies proved the role of pores in promoting osteoconduction even in porous or variously rough metallic implants [16], which are easier to process than ceramics and have eventually led to the development and clinical use of trabecular metal in orthopedics and dentistry.

Recently, porous bioactive glass coatings have also been manufactured by sponge replication [17–19] or laser cladding [20,21] on the outer surface of the alumina/zirconia monolithic cup in order to take advantage from both the trabecular bone-like morphology of the coating and the inherent bioactivity of the glass used.

The present study aims at investigating the suitability of laser treatments to impart a controlled surface texture to alumina/zirconia composite ceramics for potential use in the fabrication of acetabular components for hip joint replacement. Laser-based methods have been used in the field of bioceramics to manufacture bioactive non-porous implants with complex (curved) geometry [22–24] and coatings [25,26], but, to the best of the authors' knowledge, their potential suitability for the surface ablation of ceramic acetabular cups has not been reported yet.

2. Materials and Methods

2.1. Manufacturing of Ceramic Samples

In order to proceed to a set of experiments to test the suitability of laser treatments, it was necessary to manufacture a series of alumina/zirconia ceramic samples in the shape of acetabular cups and discs. In order to obtain alumina/zirconia shapes, some different approaches have already been explored. Whilst additive manufacturing has proven to be a suitable procedure for obtaining prosthetic parts with a controlled porosity [27], the approach via subtractive manufacturing technologies [28] is the one that, up to now, yields to the best results if high-load-bearing applications are the major goal.

For the cups required for testing, the raw material utilized was a powder composition of alumina/zirconia in a 75/25 (wt %) ratio, which undertook a process of hydrostatical compression to achieve a minimum consistency level, i.e., brown cylinders of 50 mm diameter and 100 mm of height, as shown in Figure 1a. These brown cylinders were then heat treated in a first pre-sintering process with a total duration of 17 h in an oven with forced air circulation (Hobersal Mod JM 3/16, Hobersal, Caldes de Montbui, Spain). The pre-sintering curve contained a first stop of 2 h at 100 °C to remove humidity and a final stop of 2 h at 1200 °C to achieve further mechanical stability in order to be machined, i.e., green cylinders were obtained.

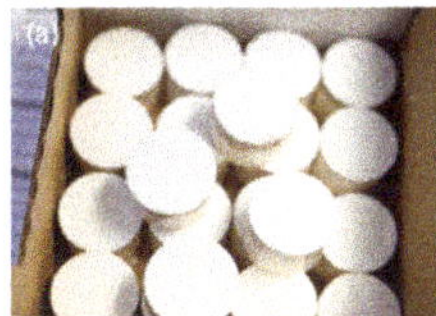

Figure 1. Raw material different stages of transformation during the manufacturing of the cup samples: (**a**) brown cylinders, (**b**) green cylinders fixed to the aluminum support discs, (**c**) green hemispheres before cross-cutting, and (**d**) sintered cups.

The subjection methods utilized for fixing the green cylinders to be machined in the milling center were those described in [29]. In particular, for the machining of the outer surface, a two-component epoxy resin was used to fix each cylinder onto an aluminum disc that was then clamped by the mill clutches, as shown in Figure 1b. Concerning the machining of the inner surface, a specific tailor-made tooling was utilized to hold each of the cylinders already processed in the form of a hemisphere, as per the shape depicted in Figure 1c.

The machining of the outer surface required up to three milling operations (end mill, ball-nose mill, and T-slot cutter operations) whilst the machining of the inner surface required two milling operations (end mill and ball-nose mill operations). Both surfaces' machining processes were performed by a vertical CNC-3 axis milling machine (Milltronics Mod RH20, Milltronics, Waconia, MN, USA). The separation of the semi-processed cylinders (hemispheres) from the aluminum discs was performed by a simple crosscut of the ceramics.

Once machined, the green cups were heat treated in a sintering process with a total duration of 30 h in an oven with forced air circulation (Hobersal Mod JM 3/16, Hobersal, Caldes de Montbui, Spain). The sintering curve contained a 2 h stop at 160 °C that yielded the final mechanical properties, i.e., ceramic cup samples to be utilized in the subsequent steps, as shown in Figure 1d.

The details of the process for obtaining the discs required for testing are fully described in [21]. Comparing with the process for the cups, all steps (hydrostatic pressing, machining, and heat treatments) were analogous, taking into consideration that the dimensions of the initial cylinders were much smaller in the discs' case. Also, the machining operations for the discs were simpler, as they could be undertaken simply via turning operations in a Multitask Centre [21].

2.2. Laser Treatment

Several structured surfaces have been generated on the alumina/zirconia ceramic discs using a Q-switched Nd:YVO4 laser source operating in near infrared (wavelength λ = 1064 nm) (Rofin PowerLine E20, Rofin, Gilching, Germany). This source provides a laser beam with the following characteristics: M2 < 1.2, about 10 ns pulse length, up to 200 kHz repetition rate, and up to 800 µJ energy per pulse. The beam was delivered with a galvoscanner head with 254 mm focal length field optics, providing a spot size of 70 µm full width at half maximum (FWHM) diameter.

The dominant regime during the process was a combination between ablation (primary effect) and melting—this was the result of smart scanning strategies used to irradiate the surface with the laser beam, provided by a Q-switched laser source of high repetition rate and pulse length in the nanosecond range. This laser beam has a Gaussian energy distribution, which means that the higher energy density is concentrated in the middle of the spot (ablative behavior).

The laser source was equipped with scanner heads having galvanometric mirrors, as shown in Figure 2. The treated surface was scanned combining parallel lines and different overlapping angles between the lines in order to produce periodic or random patterns on the surface. Figure 2 shows a scheme of the experimental set-up adopted in this study.

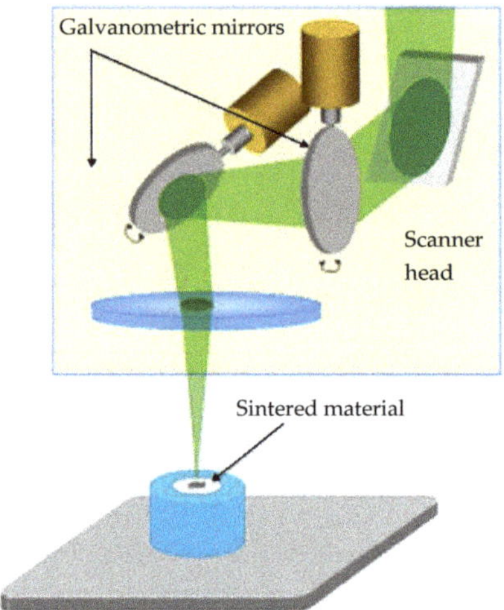

Figure 2. Laser treatment of ceramic surfaces: experimental set-up.

The main processing parameters to take into account in laser surface texturing include the laser wavelength (which is fixed as it depends on the laser source), the energy per pulse (which depends on the supplied power and the repetition rate), the scanning speed, the repetition rate (which is the number of laser pulses emitted in 1 s), the pulse overlapping (which depends on the scanning speed and the repetition rate), the repetition number (which is the number of overlapped scanning layers to generate the final pattern), the number of cycles, and the focal distance (which was fixed on the sample surface).

Before undergoing surface laser treatment, the ceramic discs were embedded in resin, cut, and polished by a diamond paste in order to reproduce the finely polished surface of currently-used real prosthetic components used in clinical applications.

Each scanning layer was generated with parallel lines separated by an interspace of 100 µm. The laser beam scanned the surface using parallel lines to cover the treated area and the various surface textures were obtained after several repetitions, as shown in Figure 3. The details of the processes are summarized in Table 1.

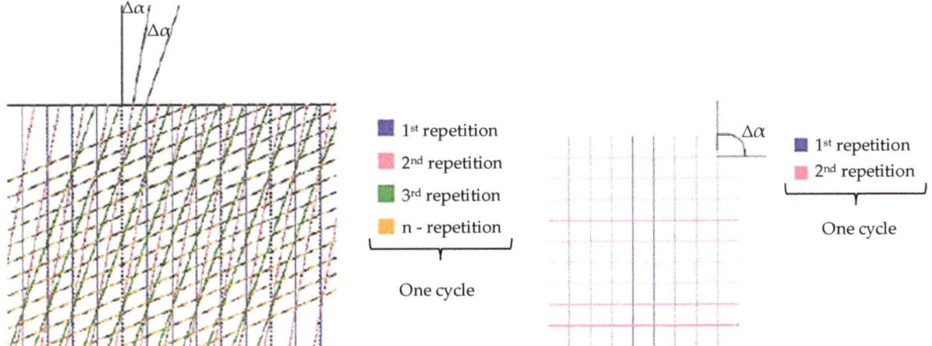

Figure 3. Laser treatment of ceramic surfaces: scheme of the process with different angles ($\Delta\alpha = 20°$ or $90°$).

Table 1. Laser process parameters used to treat the ceramic discs.

Sample Identity	Energy per Pulse (µJ)	Repetition Rate (kHz)	Scanning Speed (mm/s)	$\Delta\alpha$ (°)	Repetition Number	Cycles
1a	240	60	1000	20	20	1
2a	240	60	1000	20	20	5
1b	240	60	300	20	20	1
2b	240	60	300	20	20	5
1c	240	20	1000	20	20	1
2c	240	20	1000	20	20	5
1d	240	20	300	20	20	1
2d	240	20	300	20	20	5
3d	555	20	300	20	20	1
4a	240	60	300	90	2	1
4b	240	60	300	90	2	2
4c	240	60	300	90	2	3
4d	240	60	300	90	2	4
5	240	60	300	90	2	10

While the application of laser treatment to the surface of flat ceramic discs was relatively easy from a technological viewpoint, adapting this procedure to a ceramic cup was a challenging issue due to the need for scanning a 3D hemispheric geometry. A simple vertical scanning from the 0° position was impossible as the attach angle would vary as much as the distance from the beam focuses to the surface. Of course, there was the same problem if the cup was rotated and treated from the 90° position. In order to tackle this challenge, it was decided to divide the hemispheric cup in sectors to be processed one by one as they were considered almost flat surfaces. In this way, each of these surfaces would not have an excessive attack angle or focal distance variation. Each cup was radially divided in 72 sectors having an angular step size of 5°; four axial divisions were also considered as shown in Figure 4. The first of these divisions was laser-treated from the 90° positions while the rotary axis turned the ceramic cup of 5° per step. Then, the second and third divisions were processed using a rotatory axis of 45° according to the procedure explained previously. Finally, the ceramic cup was treated from 0° positions in its fourth division without the need for sectors or rotatory axes. Figure 4 shows a scheme of the programmed trajectory to treat the ceramic cups and Figure 5 shows the cup during laser treatment. The process parameters used to obtain a laser-textured cup prototype were the same adopted for the sample 7a.

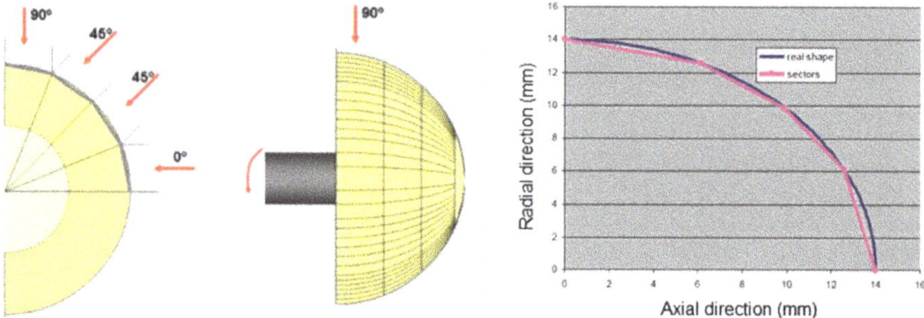

Figure 4. Schemes showing how the ceramic cup was processed by laser treatment.

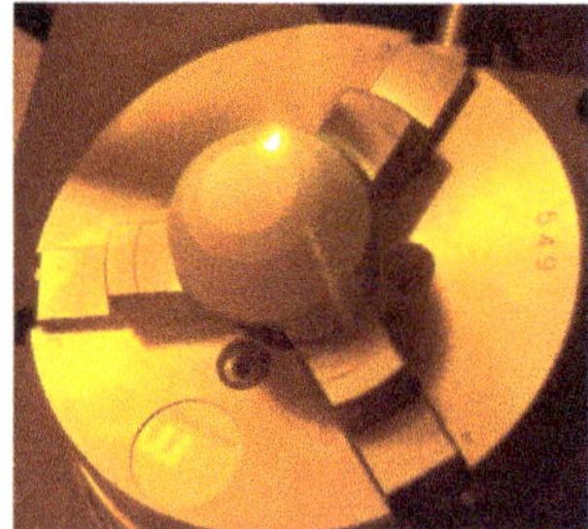

Figure 5. Camera picture showing the real-time laser texturing of the outer surface of the cup.

2.3. Characterization

2.3.1. Microstructure

Phase analysis of the ceramic samples before and after laser treatment was carried out by wide-angle X-ray diffraction (XRD; X'Pert Pro PW3040/60 diffractometer, PANalytical, Eindhoven, The Netherlands) using Cu Kα incident radiation. The XRD pattern was recorded in the 2θ-range of 10°–70° with a step size of 0.2° and a time per step of 0.5 s. Phase analysis and quantification were performed using X'Pert HighScore software (version 2.2b, PANalytical, Eindhoven, The Netherlands) equipped with the PCPDFWIN database. These XRD analyses were performed on solid sintered samples.

2.3.2. Morphology and Composition

Morphological characterization was performed using both optical and scanning electron microscopy (SEM) on selected laser-treated samples. Specifically, an Olympus metallographic optical microscope (Olympus Global, Tokyo, Japan) and a Hitachi FE-4500 field-emission SEM equipment (Hitachi Global, Tokyo, Japan) were used. Compositional analysis was performed by an energy dispersive X-ray spectroscopy (EDS) microprobe attached to the SEM. The samples were sputter-coated with an ultrathin metallic layer prior to undergoing SEM and EDS analyses.

2.3.3. Surface Roughness

Topographical analyses were carried out through the assessment of surface roughness under the ISO 4288 standard [30] using a contact profilometer (Dektak 8, Veeco, Plainview, NY, USA). Six measure lines along the transversal and longitudinal (orthogonal) directions were considered to achieve an average of the roughness value.

Amplitude parameters are the most important and commonly-used parameters to characterize surface topography. They are used to measure the vertical characteristics of the surface deviations. The arithmetic average height (R_a) is the most universally-used roughness parameter for general quality control. It is defined as the average absolute deviation of the roughness irregularities from the mean line over one sampling length. This parameter is easy to define and measure, and it gives a general description of height variations.

The surface topography can be also described by the root mean square roughness (R_q), which represents the standard deviation of the distribution of surface heights. This parameter is more sensitive than R_a to large deviations from the mean line. The R_q mean line is the line that divides the profile so that the sum of the squares of the deviations of the profile height from it is equal to zero.

R_a and R_q of each sample are expressed as mean ± standard deviation calculated on six values.

3. Results and Discussion

Figure 6 shows the XRD pattern of the sintered ceramic samples. Alumina (Al_2O_3) and tetragonal zirconia (ZrO_2) were detected and quantified in a 76-to-24 wt % ratio, which is consistent with the theoretical expectations and confirms the composite nature of the material.

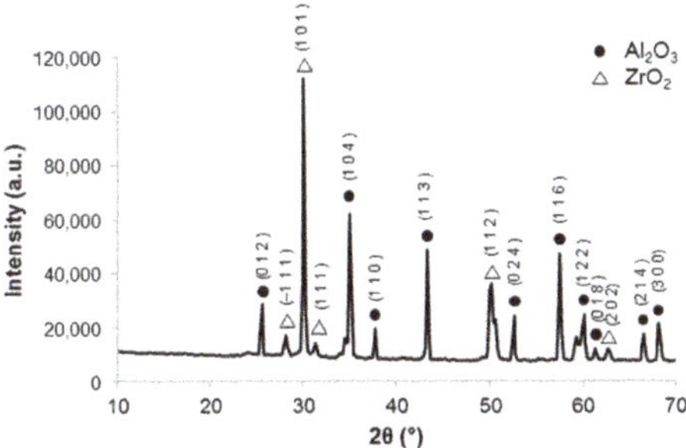

Figure 6. XRD pattern of a sintered ceramic sample.

Figure 7a reports a SEM micrograph on a polished cross-section of the sintered material, where bright inclusions seem to be embedded in a dark matrix. The compositional mapping reported in Figure 7b–d allowed the visualization of the distribution of the elements on the cross-sectional area. The distribution of colors revealed that the dark matrix corresponds to Al_2O_3 and the bright zones can be assigned to ZrO_2. These observations are complementary and consistent with the XRD results shown in Figure 6.

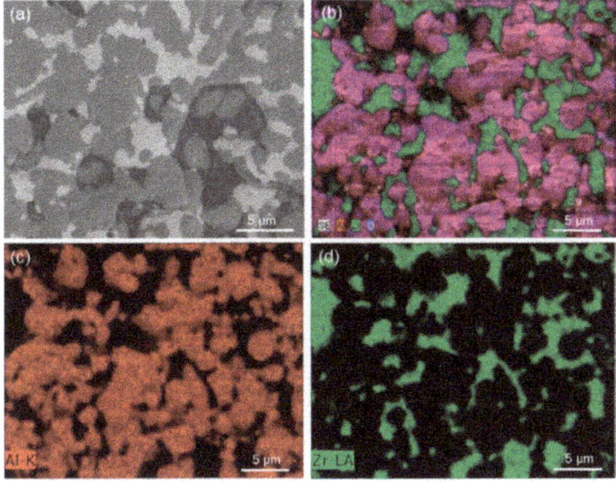

Figure 7. Morphological and compositional analyses of the sintered composite ceramics: (**a**) SEM micrograph and (**b**) compositional map achieved by energy dispersive X-ray spectroscopy (EDS), (**c**,**d**) show the distribution of Al (red) and Zr (green), respectively, with reference to the image (**b**).

Polished ceramic samples were processed by near-infrared laser under different conditions in order to impart different textures to the surface. It was generally observed that the samples acquired a dark color after laser treatment, as shown in Figure 8a. The XRD pattern of laser-treated samples (not shown) revealed the same crystalline phases (alumina and zirconia) that were detected in the original ceramic material, which however exhibited a white color. The only differences between the XRD patterns of as-such and laser-treated ceramics were the lower intensity of the peaks and the noisier pattern in the latter, which were due to the presence of the laser-induced surface roughness.

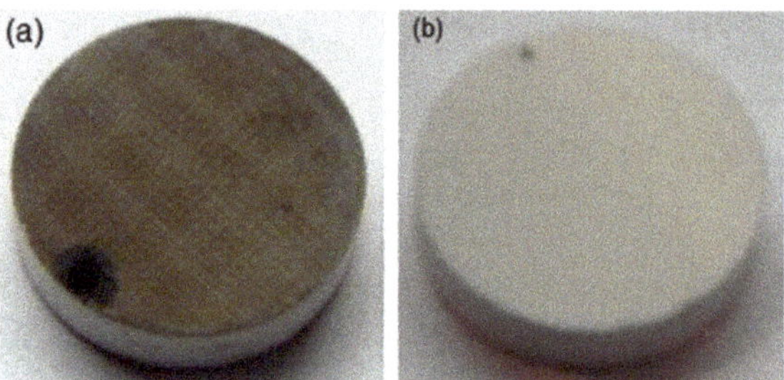

Figure 8. Laser-treated sample (**a**) before and (**b**) after post-treatment in an oven at 700 °C.

The explanation for the darkness of laser-treated samples could be related to oxygen deficiency of zirconia grains. Darkening of this ceramic material is a well-known phenomenon reported in oxygen sensors when they are exposed to reducing atmospheres. Oxygen vacancies promote the formation of stable absorbing defects, i.e., color centers in zirconia when they are irradiated with the laser beam [31,32]. This phenomenon was also confirmed by a simple experiment—after being exposed to an oxidizing atmosphere at 700 °C for 1 h, the darkened disc-shaped samples recovered their natural white color, as shown in Figure 8b.

As reported in Table 2, ceramic samples with no laser treatment exhibited a submicronic surface roughness, which was due to the minimal residual porosity after sintering; this is also consistent with the morphological observation shown in Figure 7a. The roughness profiles were dominated by valleys rather than peaks protruding from the surface, which is a key requirement in clinical applications in order to reduce the risk of wear of the adjacent prosthetic component and/or tissue abrasion [33].

Table 2. Surface roughness parameters of the ceramic samples with different laser-induced textures.

Sample Identity	R_a (µm)	R_q (µm)
As-such ceramic	0.26 ± 0.025	0.39 ± 0.038
1a	4.48 ± 0.47	6.78 ± 0.62
2a	7.25 ± 0.66	8.58 ± 1.00
1b	11.08 ± 0.64	14.10 ± 0.55
2b	31.03 ± 4.91	38.50 ± 5.88
1c	3.38 ± 0.39	4.18 ± 0.39
2c	7.75 ± 0.47	9.28 ± 0.67
1d	3.45 ± 0.29	4.4 ± 0.36
2d	5.48 ± 0.26	6.90 ± 0.26
3d	8.78 ± 1.93	12.18 ± 3.30
4a	3.78 ± 0.68	4.83 ± 0.76
4b	6.55 ± 0.41	8.08 ± 0.46
4c	8.55 ± 1.24	10.28 ± 1.51
4d	9.78 ± 1.04	12.05 ± 0.99
5	18.84 ± 0.28	21.85 ± 0.26
5 after heat treatment	18.66 ± 0.34	21.53 ± 0.37

In general, laser treatments increased the surface roughness (R_a and R_q) of the sintered ceramic samples up to two orders of magnitude depending on the processing conditions, as shown in Table 2. In general, the relatively low standard deviation for the values of roughness suggest a promising reproducibility of the sample and, thus, a good repeatability of the texturing process.

Some more specific considerations about the influence of the different parameters can also be presented. If all the other laser parameters were kept fixed, an increase of the number of cycles increased the roughness due to the higher ablation of the material surface (see comparison of samples 2a and 1a, 2b and 1b, 2c and 1c, as shown in Figure 9, and 2d and 1d). The decrease of the scanning speed increased the overlap between consecutive pulses, thereby involving an increment of the roughness (see comparison of samples 1a and 2b). The variation of the repetition rate seems to have a very limited effect on the surface roughness (comparison between samples 2a and 2c; no significant differences in terms of R_a, minimal differences in terms of R_q). An increase of roughness was also observed with increasing the energy per pulse (sample 3d vs. 1d)

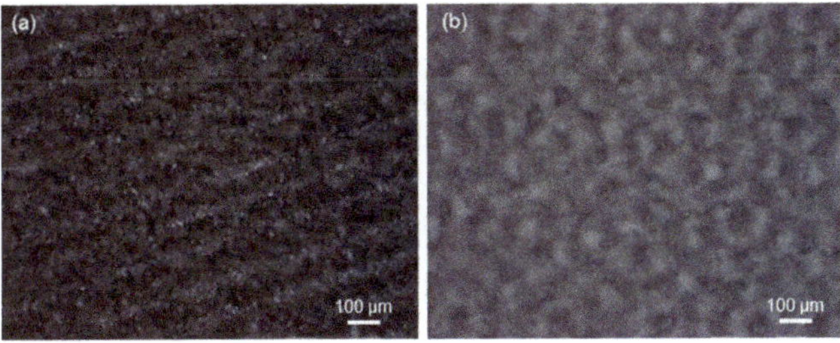

Figure 9. Optical micrographs showing the laser-treated surfaces of (**a**) sample 1c ($\Delta\alpha = 20°$, number of cycles = 1) and (**b**) sample 2c ($\Delta\alpha = 20°$, number of cycles = 5).

The orientation of the surface texture was dictated by the parameter $\Delta\alpha$, as shown in Figure 10. A symmetric textural pattern was obtained as a consequence of using $\Delta\alpha = 90°$, as shown in Figure 10a; on the contrary, a more randomly oriented roughness can be observed in Figure 10b due to an angle of 20° for the laser beam.

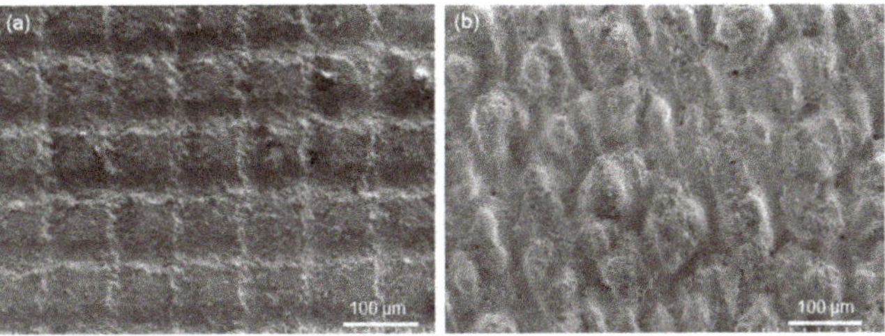

Figure 10. Effect of the parameter $\Delta\alpha$ on the texture orientation: (**a**) sample 4a ($\Delta\alpha = 90°$) and (**b**) sample 1b ($\Delta\alpha = 20°$).

As shown in Figure 11, the orientation of the surface texture was independent of the number of cycles used for the laser treatment and its regularity and periodicity was preserved.

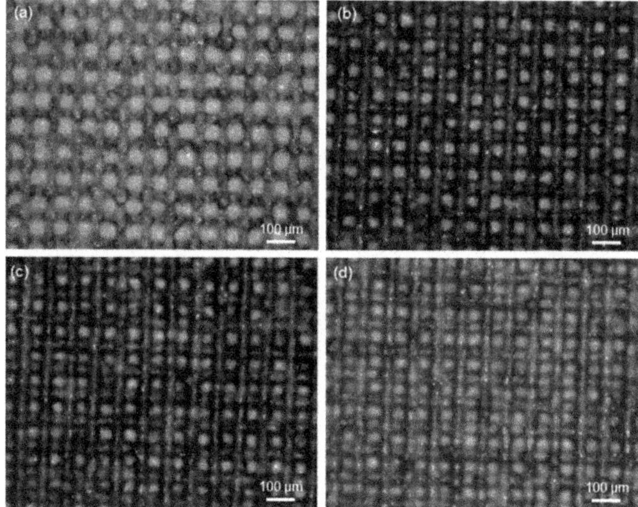

Figure 11. Optical micrographs showing the laser-treated surfaces ($\Delta\alpha = 90°$) of (**a**) sample 4a (number of cycles = 1), (**b**) sample 4b (number of cycles = 2), (**c**) sample 4c (number of cycles = 3), and (**d**) sample 4d (number of cycles = 4).

Building on the numerical results obtained in the experimentation, the significance of the effects of each of the factors investigated on the parameters R_a and R_q were analyzed. As mentioned before, the factors that were the object of study were: energy per pulse (µJ), repetition rate (kHz), scanning speed (mm/s), $\Delta\alpha$ (°), repetition number (units), and number of cycles (units). However, it should be observed that:

- Energy per pulse depends on the repetition rate; which is already in the model.
- In the set of conducted experiments, $\Delta\alpha$ and repetition number were interdependent in the different levels (i.e., in the cases where $\Delta\alpha$ was fixed as 20°, repetition number was always fixed as 20; and in all cases when $\Delta\alpha$ was fixed as 90°, repetition number was always fixed as 2.

Therefore, the screening analysis was performed with the variables repetition rate (kHz), scanning speed (mm/s), $\Delta\alpha$ (°), and number of cycles (units).

Once the conditions were fixed, the data available were analyzed in a frame in the form of a design of experiments (DOE) with four factors, two levels, fractional ($2^{4-1} = 8$ runs), one block, and no central points. This design, which is presented in Table 3, aimed at screening the significance of the factors analyzed in the overall response for R_a and R_q values.

Table 3. Design of experiments with four factors: A: repetition rate (kHz), B: scanning speed (mm/s), C: $\Delta\alpha$ (°), and D: number of cycles (units). DOE: design of experiments.

DOE Experiment Id	A: Repetition Rate (kHz)	B: Scanning Speed (mm/s)	C: $\Delta\alpha$ (°)	D: Number of Cycles (units)
1	20	300	20	1
2	60	300	20	5
3	20	1000	20	5
4	60	1000	20	1
5	20	300	90	5
6	60	300	90	1
7	20	1000	90	1
8	60	1000	90	5

The data were analyzed by making use of Minitab statistical software (Minitab R18.1, Minitab LLC, State College, PA, USA). The results showed a relatively stronger significance of factor D: number of cycles, compared to the others, as it can be identified in Figure 12a,b. Due to the limited number of experiments in this fractional design, this weak signification would require further investigation in order to decide upon the importance of the response yield. Also, the levels low (number of cycles equal to 1) and high (number of cycles equal to 5) of factor D: number of cycles were not taking into account the entire range of experimentation, but only the one with enough data for the construction of the DOE framework, so further analysis should also address the bigger range.

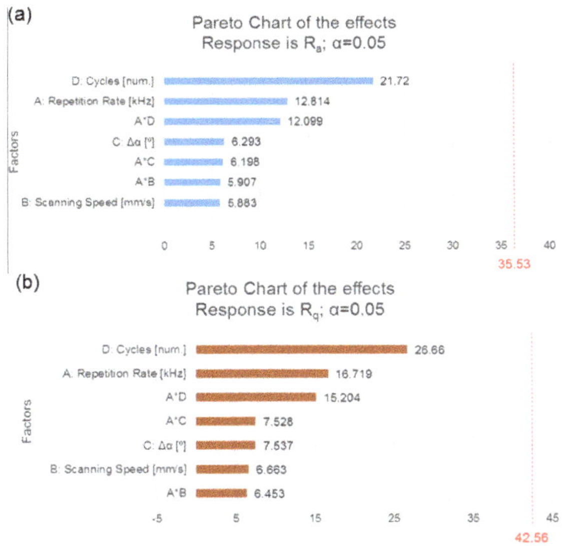

Figure 12. Pareto diagram for the significance of the effects of the variables in the DOE regarding R_a (**a**) and R_q (**b**). Variable D: number of cycles shows a relatively stronger significance than the others in both cases.

For the separated investigation on the influence of variable D: number of cycles to the level of R_a and R_q observed, all available data were used, ranging from a number of cycles equal to 1 to a number of cycles equal to 10. Indeed, a strong linear correlation was found between the number of cycles and the values of R_a and R_q, as revealed by the high value of the coefficients of determination R^2 (above 0.98 in R_a and above 0.96 in R_q), as shown in Figure 13. The resulting linear interpolating functions can be useful from a technological viewpoint in order to predict, at the design stage, the expected roughness depending on laser processing parameters.

Sample 5 exhibited an ordered textural arrangement with regular periodicity of squared "hills" and valleys created by the laser beam, as shown in Figure 14a,b. SEM investigations revealed that a two-level texture can be observed on these samples, i.e., a laser-induced roughness with R_a and R_q values around 20 µm, as shown in Table 2, combined with a finer micro-roughness on the top of the hills, where the ceramic particles were joined and partially melted together during sintering, assuming a quite rounded morphology, as shown in Figure 14c. It is known that nanoparticles can form during processes of laser ablation [34,35]; however, the size of these nanoparticles was shown to typically be within the range of few nanometers, hence they are too small to affect the quality of the surface texture at the scale investigated in the present study. The hierarchical roughness of the samples is expected to be highly favorable to promote osteointegration of the ceramic implants. In fact, the surface roughness due to laser ablation is comparable to the size of human osteoblastic cells, which exhibit a typical diameter within 15–30 µm [36] and can then potentially enter the valleys created by the laser, thereby leading to new bone tissue interlocking within the implant surface. Furthermore, the fine roughness on the top of the hills could also play a role in promoting bone cell behavior towards paths of osteogenesis and self-repair. Textural properties of implant surfaces are known to greatly influence cell responses in vitro and in vivo [37]. The early studies carried out by Schwartz and Boyan [38] gave a first evidence that osteoblasts attach and spread preferably on implant surfaces with a diffused micrometric roughness. The micrometric and nanometric peaks and valleys of implant surfaces were shown to affect the organization of bone cell cytoskeleton and the intracellular transduction signaling pathways, thus emphasizing the bone regenerative capacity [39,40].

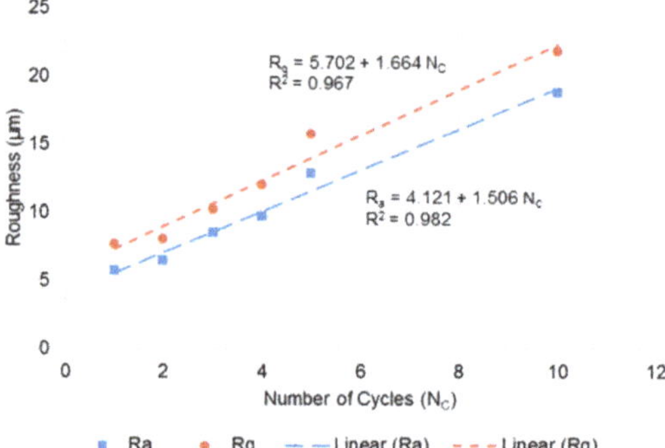

Figure 13. Relationship between surface roughness (R_a and R_q) and number of cycles used during the laser treatment.

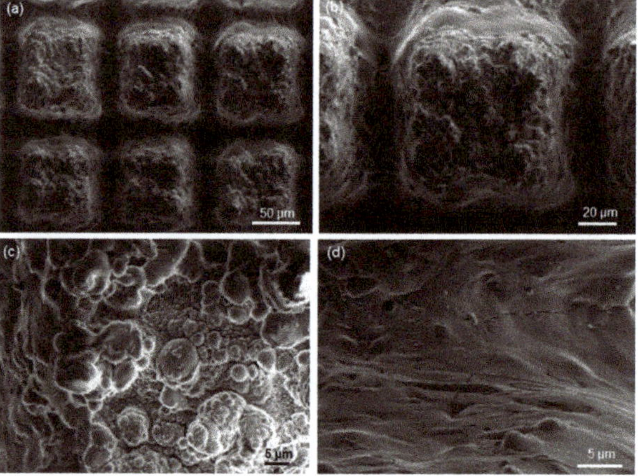

Figure 14. SEM micrographs of sample 5: (**a**) general view and (**b**) detail of the surface texture, (**c**) top surface of a hill and (**d**) detail of the area between two adjacent hills.

Comparison of Figure 14c and d reveal the morphological difference that exists between the top of the hills and the bottom of the laser-created valleys; specifically, Figure 14d shows that melting of the ceramic surfaces produces small cracks in the material as a consequence of the high cooling rate.

As shown in Table 2, the heat post-treatment carried out to eliminate the dark color of the samples caused no significant alteration in the values of R_a and R_q.

A laser-textured prototype of the ceramic cup with square-shaped roughness was successfully obtained, as shown in Figure 15. The laser parameters used to process sample 5 were adopted for this purpose, under the appropriate technological optimization. This choice was based on the values of roughness reported in Table 2; except for sample 5, all the other laser-textured specimens had a surface roughness above (sample 2b) or below (all the other cases) the typical size range of osteoblastic cells (15–30 μm). Also, in the case of the surface-textured cup, a heat treatment in oxidizing atmosphere was suitable to recover the original white appearance of the cup, as shown in Figure 15b.

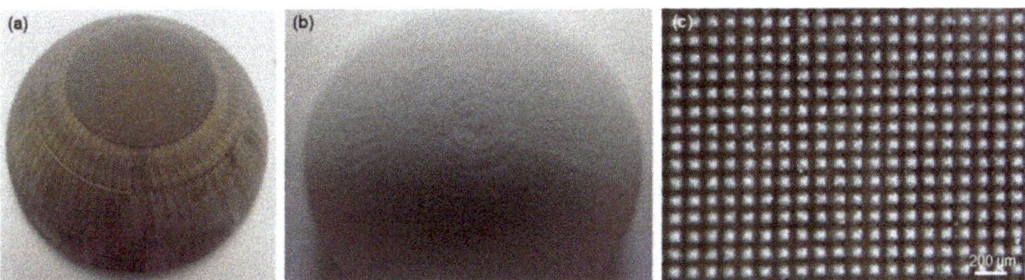

Figure 15. Laser-textured ceramic cup: (**a**) sample after laser treatment, (**b**) sample after thermal treatment in an oven at 700 °C, (**c**) optical micrograph of the surface.

4. Conclusions

In this work, different surface textures were imparted to alumina/zirconia ceramic samples by applying infrared laser ablation strategies. Surface roughness could be modulated in the range of 3 to 30 μm by varying the laser process parameters (e.g., number of cycles, scanning speed, energy per pulse). In general, the roughness increased with (i) increasing number of cycles due to the higher ablation of the material, (ii) decreasing scanning speed, or (iii) increasing the energy per pulse. Critical analysis based on DOE suggests the relatively stronger significance of the number of cycles compared to the other factors. Furthermore, highly-arranged or random-like textures could be obtained depending on the variation of the laser process angle. Morphological investigations revealed a hierarchical texture formed by a laser-induced roughness combined with a finer micro-roughness deriving from the sintering process of the ceramic bodies. The method proposed for laser surface texturing of flat ceramic samples was properly adapted for application on real acetabular cups for hip joint prosthesis, and a prototype was successfully obtained. The produced surface roughness is expected to be suitable for promoting bone tissue in-growth and interlocking in acetabular prosthetic implants; future in vivo studies will allow elucidating if a renaissance of surface-textured cementless monolithic ceramic cups is actually possible.

Author Contributions: Conceptualization, F.B., J.M.-C., M.A.M. and C.V.-B.; Methodology, F.B., J.M.-C., M.A.M. and C.V.-B.; Software, F.B., J.M.-C. and M.A.M.; Validation, C.V.-B.; Formal Analysis, F.B.; Investigation, F.B., J.M.-C., M.A.M. and C.V.-B.; Resources, F.B., J.M.-C., M.A.M. and C.V.-B.; Data Curation, F.B. and M.A.M.; Writing—Original Draft Preparation, F.B. and J.M.-C.; Writing—Review and Editing, F.B., J.M.-C., M.A.M. and C.V.-B.; Visualization, F.B. and J.M.-C.; Supervision, C.V.-B.; Project Administration, C.V.-B. and J.M.-C.; Funding Acquisition, C.V.-B. and J.M.-C.

Funding: This research was funded by the European Commission, Research Executive Agency, EUFP7/2007-2013 SME-2011-1, Specific Programme "Capacities": Research for the benefit of SMEs (BSG-SME), under grant agreement no. 286548 ("Monoblock acetabular cup with trabecular-like coating"—MATCh project).

Conflicts of Interest: The authors declare no conflict of interest. The funders had no role in the design of the study; in the collection, analyses, or interpretation of data; in the writing of the manuscript, or in the decision to publish the results.

References

1. López-López, J.A.; Humphriss, R.L.; Beswick, A.D.; Thom, H.H.Z.; Hunt, L.P.; Burston, A.; Fawsitt, C.G.; Hollingworth, W.; Higgins, J.P.T.; Welton, N.J.; et al. Choice of implant combinations in total hip replacement: Systematic review and network meta-analysis. *BMJ* **2017**, *359*, 4651. [CrossRef] [PubMed]
2. Clegg, A.; Young, J.; Iliffe, S.; Rikkert, M.O.; Rockwood, K. Frailty in elderly people. *Lancet* **2013**, *381*, 752–762. [CrossRef]

3. Edwards, L.D.; Levin, S. Complications from total hip replacement with the use of acrylic cement. *Health Serv. Rep.* **1973**, *88*, 857–867. [CrossRef] [PubMed]

4. Latham, B.; Goswami, T. Effect of geometric parameters in the design of hip implants-paper IV. *Mater. Des.* **2004**, *25*, 715–722. [CrossRef]

5. Sargeant, A.; Goswami, T. Hip implants: Paper V. Physiological effects. *Mater. Des.* **2006**, *27*, 287–307. [CrossRef]

6. Kharmanda, G. Reliability analysis for cementless hip prosthesis using a new optimized formulation of yield stress against elasticity modulus relationship. *Mater. Des.* **2015**, *65*, 496–504. [CrossRef]

7. Rahaman, M.N.; Yao, A.; Sonny Bal, B.; Garino, J.P.; Ries, N.D. Ceramics for prosthetic hip and knee joint replacement. *J. Am. Ceram. Soc.* **2007**, *90*, 1965–1988. [CrossRef]

8. O'Leary, J.F.; Mallory, T.H.; Kraus, T.J.; Lombardi, A.V., Jr.; Lye, C.L. Mittelmeier ceramic total hip arthroplasty. A retrospective study. *J. Arthroplast.* **1988**, *3*, 87–96. [CrossRef]

9. Griss, P.; Claus, A.; Scheller, G. Analyse Unserer Erfahrungen Mit Keramik/Keramik-Huftendoprothesen der Ersten Generation. In *Reliability and Long-Term Results of Ceramics in Orthopaedics*; Sedel, L., Willmann, G., Eds.; Thieme: Stuttgart, Germany, 1999; pp. 43–47.

10. Gierse, H.; Maaz, B.; Hofer, C.; Gruner, S. The ceramic cup type Lindenhof. Results 10–14 years after implantation. *Arch. Orthop. Trauma Surg.* **1996**, *115*, 167–170. [CrossRef]

11. Garcia-Cimbrelo, E.; Martinez-Sayanes, J.M.; Minuesa, A.; Munuera, L. Mittelmeier ceramic-ceramic prosthesis after 10 years. *J. Arthroplast.* **1996**, *11*, 773–778. [CrossRef]

12. Rosner, B.I.; Postak, P.D.; Greenwald, A.S. Cup/liner conformity of modular acetabular designs. *Orthop. Trans.* **1995**, *19*, 469–470.

13. Schreiner, U.; Schulze, A.; Scheller, G.; Apruzzese, C.; Schwarz, M.L. Osseointegration of ceramic cement-free acetabular cups. *Z. Orthop. Unfallchir.* **2011**, *150*, 32–39. [CrossRef] [PubMed]

14. Forgon, M.; Mammel, E.; Trombitas, K.; Kacsalova, L.; Draveczki, I. Morphological investigations of a porous aluminium oxide ceramic and the consequences for clinical application. *Arch. Orthop. Trauma Surg.* **1987**, *106*, 385–389. [CrossRef] [PubMed]

15. Cornell, N.C.; Lane, J.M. Current understanding of osteoconduction in bone regeneration. *Clin. Orthop.* **1998**, *355*, 267–273. [CrossRef] [PubMed]

16. Spriano, S.; Yamaguchi, S.; Baino, F.; Ferraris, S. A critical review of multifunctional titanium surfaces: New frontiers for improving osseointegration and host response, avoiding bacteria contamination. *Acta Biomater.* **2018**, *79*, 1–22. [CrossRef]

17. Baino, F.; Minguella, J.; Kirk, N.; Montealegre, M.A.; Fiaschi, C.; Korkusuz, F.; Orlygsson, G.; Vitale-Brovarone, C. Novel full-ceramic monoblock acetabular cup with a bioactive trabecular coating: Design, fabrication and characterization. *Ceram. Int.* **2016**, *42*, 6833–6845. [CrossRef]

18. Baino, F.; Vitale-Brovarone, C. Trabecular coating on curved alumina substrates using a novel bioactive and strong glass-ceramic. *Biomed. Glasses* **2015**, *1*, 31–40. [CrossRef]

19. Baino, F.; Tallia, F.; Novajra, G.; Minguella-Canela, J.; Montealegre, M.; Korkusuz, F.; Vitale-Brovarone, C. Novel Bone-Like Porous Glass Coatings on Al_2O_3 Prosthetic Substrates. *Key Eng. Mater.* **2014**, *631*, 236–240. [CrossRef]

20. Baino, F.; Montealegre, M.A.; Orlygsson, G.; Novajra, G.; Vitale-Brovarone, C. Bioactive glass coatings fabricated by laser cladding on ceramic acetabular cups: A proof-of-concept study. *J. Mater. Sci.* **2017**, *52*, 9115–9128. [CrossRef]

21. Baino, F.; Minguella-Canela, J.; Korkusuz, F.; Korkusuz, P.; Kankılıç, B.; Montealegre, M.A.; De los Santos-López, M.A.; Vitale-Brovarone, C. In vitro assessment of bioactive glass coatings on alumina/zirconia composite implants for potential use in prosthetic application. *Int. J. Mol. Sci.* **2019**, *20*, 722–737. [CrossRef]

22. Comesaña, R.; Lusquiños, F.; Del Val, J.; Malot, T.; López-Álvarez, M.; Riveiro, A.; Quintero, F.; Boutinguiza, M.; Aubry, P.; De Carlos, A.; et al. Calcium phosphate grafts produced by rapid prototyping based on laser cladding. *J. Eur. Ceram. Soc.* **2011**, *31*, 29–41. [CrossRef]

23. Comesaña, R.; Lusquiños, F.; Del Val, J.; López-Álvarez, M.; Quintero, F.; Riveiro, A.; Boutinguiza, M.; De Carlos, A.; Jones, J.R.; Hill, R.G.; et al. Three-dimensional bioactive glass implants fabricated by rapid prototyping based on CO_2 laser cladding. *Acta Biomater.* **2011**, *7*, 3476–3487. [CrossRef] [PubMed]

24. Del Val, J.; López-Cancelos, R.; Riveiro, A.; Badaoui, A.; Lusquiños, F.; Quintero, F.; Comesaña, R.; Boutinguiza, M.; Pou, J. On the fabrication of bioactive glass implants for bone regeneration by laser assisted rapid prototyping based on laser cladding. *Ceram. Int.* **2016**, *42*, 2021–2035. [CrossRef]

25. Lusquiños, F.; Pou, J.; Boutinguiza, M.; Quintero, F.; Soto, R.; León, B.; Pérez-Amor, M. Main characteristics of calcium phosphate coatings obtained by laser cladding. *Appl. Surf. Sci.* **2005**, *247*, 486–492. [CrossRef]

26. Comesaña, R.; Quintero, F.; Lusquiños, F.; Pascual, M.J.; Boutinguiza, M.; Durán, A.; Pou, J. Laser cladding of bioactive glass coatings. *Acta Biomater.* **2010**, *6*, 953–961. [CrossRef] [PubMed]

27. Minguella-Canela, J.; Villegas, M.; Poll, B.; Tena, G.; Ginebra, M.P. Automatic casting of advanced technical ceramic parts via open source high resolution 3D printing machines. *Key Eng. Mater.* **2014**, *631*, 269–274. [CrossRef]

28. Ayats, J.R.G.; Canela, J.M. Development of a methodology for the materialisation of ceramic rapid prototypes based on substractive methods. *Arch. Mater. Sci.* **2007**, *28*, 9–14.

29. Minguella-Canela, J.; Cuiñas, D.; Rodríguez, J.V.; Vivancos, J. Advanced manufacturing of ceramics for biomedical applications: Subjection methods for biocompatible materials. *Procedia Eng.* **2013**, *63*, 218–224. [CrossRef]

30. *Geometrical Product Specifications (GPS)—Surface Texture: Profile Method—Rules and Procedures for the Assessment of Surface Texture*; ISO 4288; International Organization for Standardization: Geneva, Switzerland, 1996.

31. Orera, V.M.; Merino, R.I.; Chen, Y.; Cases, R.; Alonso, P.J. Intrinsic electron and hole defects in stabilized zirconia single crystals. *Phys. Rev. B* **1990**, *42*, 9782–9789. [CrossRef]

32. Orera, V.M.; Merino, R.I.; Chen, Y.; Cases, R.; Alonso, P.J. Electron and hole trapped defects produced by thermo-reduction or irradiation in stabilized zirconia. *Radiat. Eff. Defects. Solids* **1991**, *119*, 907–912. [CrossRef]

33. Baino, F.; Gautier di Confiengo, G.; Faga, M.G. Fabrication and morphological characterization of glass-ceramic orbital implants. *Int. J. Appl. Ceram. Technol.* **2018**, *15*, 884–891. [CrossRef]

34. Semaltianos, N.G. Nanoparticles by laser ablation. *Crit. Rev. Solid State Mater. Sci.* **2010**, *35*, 105–124. [CrossRef]

35. Kim, M.; Osone, S.; Kim, T.; Higashi, H.; Seto, T. Synthesis of nanoparticles by laser ablation: A review. *KONA Powder Part. J.* **2017**, *34*, 80–90. [CrossRef]

36. Reznikov, N.; Shahar, R.; Weiner, S. Bone hierarchical structure in three dimensions. *Acta Biomater.* **2014**, *10*, 3815–3826. [CrossRef] [PubMed]

37. Feller, L.; Jadwat, Y.; Khammissa, R.A.G.; Meyerov, R.; Schechter, I.; Lemmer, J. Cellular responses evoked by different surface characteristics of intraosseous Ti implants. *BioMed Res. Int.* **2015**, *2015*, 171945. [CrossRef] [PubMed]

38. Schwartz, Z.; Boyan, B.D. Underlying mechanisms at the bone-biomaterial interface. *J. Cell. Biochem.* **1994**, *56*, 340–347. [CrossRef] [PubMed]

39. Qu, Z.; Rausch-Fan, X.; Wieland, M.; Matejka, M.; Schedle, A. The initial attachment and subsequent behavior regulation of osteoblasts by dental implant surface modification. *J. Biomed. Mater. Res. A* **2007**, *82*, 658–668. [CrossRef]

40. Anselme, K.; Davidson, P.; Popa, A.M.; Giazzon, M.; Liley, M.; Ploux, L. The interactions of cells and bacteria with surfaces structured at the nanometer scale. *Acta Biomater.* **2010**, *6*, 3824–12846. [CrossRef]

 © 2019 by the authors. Licensee MDPI, Basel, Switzerland. This article is an open access article distributed under the terms and conditions of the Creative Commons Attribution (CC BY) license (http://creativecommons.org/licenses/by/4.0/).

 *coatings*

Article

Influence of Aluminum Laser Ablation on Interfacial Thermal Transfer and Joint Quality of Laser Welded Aluminum–Polyamide Assemblies

Adham Al-Sayyad [1,*], **Julien Bardon** [2], **Pierre Hirchenhahn** [3], **Regis Vaudémont** [2], **Laurent Houssiau** [3] **and Peter Plapper** [1]

[1]   Faculty of Science, Technology and Communication (FSTC), University of Luxembourg, L1359 Luxembourg, Luxembourg; peter.plapper@uni.lu

[2]   Materials Research and Technology (MRT), Luxembourg Institute of Science and Technology, L4362 Esch-sur-Alzette, Luxembourg; julien.bardon@list.lu (J.B.); regis.vaudemont@list.lu (R.V.)

[3]   Namur Institute of Structured Materials (NISM), LISE laboratory, University of Namur, 5000 Namur, Belgium; pierre.hirchenhahn@unamur.be (P.H.); laurent.houssiau@unamur.be (L.H.)

*   Correspondence: adham.alsayyad@uni.lu; Tel.: +352-466-644-6034

Received: 29 September 2019; Accepted: 12 November 2019; Published: 19 November 2019

Abstract: Laser assisted metal–polymer joining (LAMP) is a novel assembly process for the development of hybrid lightweight products with customized properties. It was already demonstrated that laser ablation of aluminum alloy Al1050 (Al) prior to joining with polyamide 6.6 (PA) has significant influence on the joint quality, manifested in the joint area. However, profound understanding of the factors affecting the joint quality was missing. This work investigates the effects of laser ablation on the surface properties of Al, discusses their corresponding impact on the interfacial thermal transfer between the joining partners, and evaluates their effects on the joint quality. Samples ablated with different parameters, resulting in a range from low- to high-quality joints, were selected, and their surface properties were analyzed by using 2D profilometry, X-ray photoelectron spectroscopy (XPS), scanning electron microscope (SEM), and energy-dispersive X-ray spectroscopy (EDX). In order to analyze the effects of laser ablation parameters on the interfacial thermal transfer between metal and polymer, a model two-layered system was analyzed, using laser flash analysis (LFA), and the thermal contact resistance (TCR) was quantified. Results indicate a strong influence of laser-ablation parameters on the surface structural and morphological properties, influencing the thermal transfer during the laser welding process, thus affecting the joint quality and its resistance to shear load.

Keywords: laser welding; metal–polymer; laser ablation; thermal contact resistance

1. Introduction

Joining metals to polymers has gained prominent interest among researchers and industries because of its ability to produce lightweight hybrid structures with tailored properties. Conventional metal–polymer joining methods, such as adhesive bonding, mechanical fastening, friction stir welding, and ultrasonic joining, have their drawbacks, as they either require high processing time, involve hazardous chemicals, cause excessive tool wear, involve geometrical constrains, or require the addition of weight to the component. The thermal joining of metals to polymers is challenging because of the significant difference of the thermal properties and melting temperatures of both materials. However, Laser-Assisted Metal–Polymer joining (LAMP) provides the ability to precisely control the energy input into the materials, giving the opportunity to thoroughly melt the polymer at the interface of the joint, while, at the same time, avoiding its degradation. In addition, LAMP has its advantages over conventional joining methods in being rapid, autogenous, and a non-contact process that offers design flexibility, along with the ability to produce miniaturized joints and minimize overlapping dimensions.

So far, research in LAMP has shown the reliability of the laser welding process in a variety of material combinations [1–4]. Preliminary surface treatments before welding have shown a significant impact on the joint strength and quality. Researchers [3–12] reported the effects of several surface pretreatments on LAMP, including mechanical, chemical, electrochemical, and laser pretreatments for metals, as well as plasma and UV ozone pretreatment for polymers. Similar to adhesive bonding, two main factors were reported in the literature as affecting the LAMP joint properties: mechanical interlocking and physicochemical bonding. However, although LAMP is a thermal joining process, the effect of surface pretreatments on the thermal transfer between the joining partners during the laser welding process, and on the joint quality, has not been investigated.

Increased surface roughness has shown to have a positive effect on enhancing LAMP joint strength by increasing mechanical interlocking effects [9,10,12]. Furthermore, an increased surface roughness might increase the surface wettability of treated metal, which in turns allows a better wetting of the molten polymer during the laser welding process. However, an increase in surface roughness results in an increase in the thermal contact resistance (TCR) across the interface of the two solid materials in contact [13,14]. When two solid surfaces come into contact, the flow of heat across the interface is mainly governed by solid-to-solid conduction at the points of contact, and conduction through the fluid occupying the noncontact area, resulting in restrictions to the heat flow [15]. Yovanovich [16] summarizes forty years of research in the field of thermal contact resistance, and in particular introduces the development of a geometric–mechanical–thermal model called the Cooper–Mikic–Yovanovich (CMY) model to predict the thermal contact resistance of conforming rough surfaces.

Researchers [17–22] developed several theoretical models to calculate thermal contact resistance. Three mechanical models—elastic, plastic, or elastic–plastic deformation of the surface asperities— were considered, assuming that surface asperities follow Gaussian height distributions about a mean plane passing through each surface, and assuming that surface asperities are randomly distributed over the apparent contact area. Yovanovich [23] summarized the TCR model developed by Cooper et al. [20] and proposed compact expression to calculate the TCR between two nominally flat solid surfaces (1 and 2) in contact assuming plastic deformation of asperities as given by Equation (1). Given that σ_1, m_1, k_1 are properties of material 1, and σ_2, m_2, k_2 are properties of material 2, $\sigma = (\sigma_1{}^2 + \sigma_2{}^2)^{0.5}$ is the effective root mean square roughness (RMS) so that $\sigma_{1,2} = \sqrt{\frac{1}{L} \int_0^L y^2(x)\mathrm{d}x}$, where L is the profile traced length; $m = (m_1{}^2 + m_2{}^2)^{0.5}$ is the effective mean asperity slope, and $m_{1,2} = \frac{1}{L} \int_0^L \left| \frac{\mathrm{d}y(x)}{\mathrm{d}x} \right| \mathrm{d}x$ is the mean absolute asperity slope. H is the microhardness of the softer material, P is the applied pressure, and k_h is the harmonic mean thermal conductivity at the interface where $k_\mathrm{h} = 2k_1 k_2/(k_1 + k_2)$. It is clear from Equation (1), which can be utilized in describing the TCR between metal and polymer in their solid state, prior to the laser joining process [24], that TCR has a proportional relation to the ratio of the root mean square roughness σ to the asperities slope, m.

$$TCR = \frac{0.8\sigma}{mk_h}\left(\frac{H}{P}\right)^{0.95} \tag{1}$$

Laser ablation was proven to be an effective and rapid surface pretreatment technique for aluminum (Al 1050) to enhance the bonding strength when welded with polyamide (PA 6.6) [1,5,25]. It was already demonstrated that laser-ablation parameters have a strong impact on the joint quality, demonstrated by the joint area and its resistance to failure, but no effect on the joint strength, i.e., its stress to failure [5]. Results have shown the prominence of cohesive failure mode indicating that interfacial adhesion is not the only factor influencing the joint quality. Preliminary results showed that the joint resistance to failure is influenced by the topography of the ablated aluminum surface, in particular by topography parameters representing the density of peaks on the surface. From the knowledge of TCR models quoted before, it is therefore hypothesized that laser-ablation parameters affect the thermal transfer between the joining partners, which would reflect on the joint area and quality, impacting its resistance to shear load. This research aims at understanding the effects of surface properties on the interfacial thermal transfer between laser ablated Al 1050 joined to PA 6.6 using laser-beam welding. It correlates the influence of surface topography and the chemistry of aluminum modified layer to the TCR and reports their consecutive effects on the joint quality. To the best of the authors' knowledge, this is the first time that the prominence of the TCR for the welding quality is evidenced in LAMP.

2. Experimental Method

2.1. Materials

In those experiments, 0.5 mm thick EN-AW1050A aluminum (Al) in half-hard state, with geometry of 30 mm × 60 mm, and 4 thick mm polyamide 6.6 (PA) purchased from Dutec (Ahaus, Germany), with the dimensions of 25 mm × 75 mm were used. Prior to the joining process, Al samples were prepared by laser ablation, and PA samples were wiped with ethanol, to remove potential surface contaminants.

2.2. Laser Ablation

Al surfaces were ablated, using a short-pulsed (ns) Nd:YVO4 laser (TruMark 6130 from TRUMPF, Ditzingen, Germany), with a wavelength of 1064 nm and spot size of 45 µm. Al-Sayyad et al. [5] already demonstrated that, from seven different ablation parameters, namely pulse frequency (f_p), beam guidance speed (V), lines, focal position, Al rolling direction, ablation hatching orientation, and laser beam power percentage, only f_p and V had a significant influence on laser welded Al–PA joints' resistance to shear failure. This research focuses on further evaluating the effects of those significant laser-ablation conditions, as shown in Table 1, on Al surface properties, thermal transfer across the joining partners, and corresponding joint quality. Based on previous investigations [5], six ablation conditions (see Table 1) were chosen for this research, as they resulted in a wide range of joint quality. Since ablation was performed with a q-switched laser, increasing pulse frequency results in decreasing the peak pulse power and fluence of the laser beam. However, overlap ratio between consecutive laser pulses (see Figure 1) depends on both pulse frequency and beam guidance speed, as described by Equation (2).

$$O\ (\%) = 100 \times \left(1 - \frac{V\left(\frac{mm}{s}\right)}{f_p\,(\text{Hz}) \times O(mm)} \right) \tag{2}$$

Table 1. Laser-ablation parameters and attributes.

Parameter	Frequency (kHz)	Speed (mm/s)	Peak Pulse Power (kW)	Fluence (J/cm^2)	Overlap Ratio (%)
P1	85	250	11	15.2	93
P2	40	1000	35	28.6	44
P3	70	1000	15	17.9	68
P4	85	1750	11	15.2	54
P5	120	1750	5	9.12	68
P5-95%	Same as Al_5 but with 95% power		4.75	8.66	68

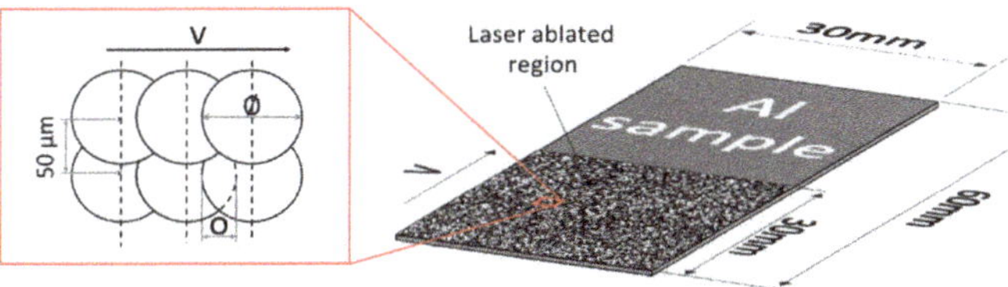

Figure 1. Schematic drawing illustrating six laser pulses irradiating Al surface during laser-ablation process and demonstrating laser pulses overlap ratio (O) between two consecutive pulses.

2.3. Laser-Beam Welding

Laser welding was performed with a fiber laser (TruFiber 400 from TRUMPF), with a wavelength of 1070 nm and a calculated spot size of 58 μm, irradiated on the Al surface after clamping the parts in an overlap configuration, as shown in Figure 2. A peak pulse power of 400 W was modulated with a pulse frequency of 25 kH and pulse duration of 35 μs. The laser beam followed a spiral trajectory, with a feed velocity (v_f) of 88.8 mm/s. Part of the irradiated laser-beam energy gets absorbed, converted to heat energy, conducted through the Al, and transferred to melt PA, thereby joining Al to PA. However, heat transfer across the interface depends on the TCR. Figure 3 shows a simplified illustration of the hypothesized thermal-energy transfer of the irradiated laser beam across the interface in the case of (a) ideal flat surfaces and (b) across rough ablated Al surfaces in contact with smooth PA modeling "real life" conditions. The presence of surface asperities at the interface causes the thermal transfer by conduction to take place mostly at the points of contact [16], thereby increasing TCR and reducing the thermal energy transfer across the interface. Single lap shear tests were performed by using Z010 from Zwick/Roell (Ulm, Germany). Samples were clamped in a vertical alignment, with a jaw-to-jaw distance of 75 mm. The crosshead pulling speed was set at 2.21 mm/min.

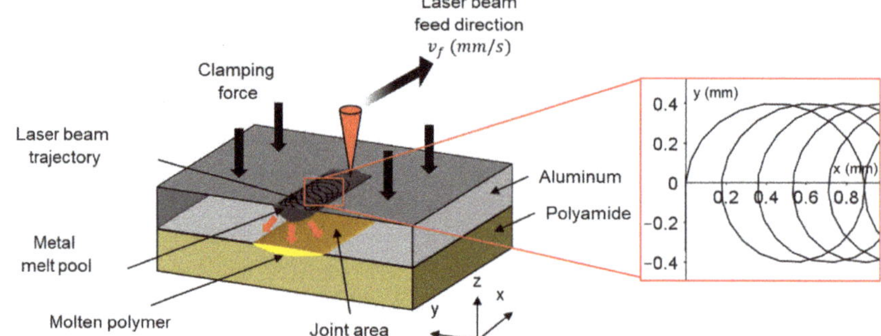

Figure 2. Schematic drawing of laser-beam-welding setup.

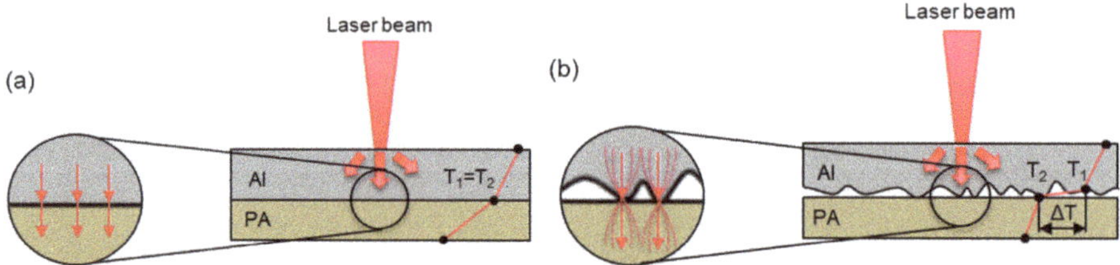

Figure 3. Schematic of hypothesized interfacial thermal-energy transfer during laser-welding process across (**a**) ideally flat conforming surfaces and (**b**) ablated aluminum surfaces in contact with smooth polyamide.

2.4. Joint-Area Assessment

In order to quantify the joint area, a dedicated experimental approach was developed. First, macroscopic images of the joint interface after failure were obtained by using a digital FUJIFILM X-Pro2 camera (Tokyo, Japan). Next, the dimensions of a single pixel were measured, using GIMP software (2.10), by correlating to a predefined scale positioned on the sample close to the weld zone. Then, the coordinates outlining the joint area were allocated, and the joint area was measured by counting the number of pixels and correlating them to the measured pixel dimension. Polarized and stitched microscopic images of the joint area were used to confirm the measurements.

2.5. X-Ray Photoelectron Spectroscopy (XPS)

The atomic composition of treated surfaces and its chemical-bonding states were investigated, using a K-alpha from Thermo Scientific (Waltham, MA, USA). An X-ray beam (Al Kα, 1486.6 eV) with a spot size of 300 μm was used in the analysis. Six regions of 1 cm^2 area were ablated on an Al sample, each region with different ablation condition (see Table 1). Six points per ablated region were investigated by measuring a survey spectrum (3 scans, 200 eV energy pass) and high-resolution spectra for the regions of Al 2p, O 1s, and C 1s (20 scans, 20 eV pass energy) atoms.

2.6. Scanning Electron Microscope (SEM)

A pressure-controlled FEI Quanta FEG 200 scanning electron microscope (SEM) from FEI Company (Hillsboro, OR, USA) was used in secondary electron mode, in order to get information about the samples' morphology. The acceleration voltage was generated at 15 kV. Six regions of 1 cm^2 area were ablated on an Al sample. The sample conductivity was enhanced by depositing a fine layer of conductive lacquer in contact with the untreated aluminum part of the sample. The area that was coated by this lacquer was not observed.

2.7. Energy-Dispersive X-Ray Spectroscopy (EDX)

EDX spectra were obtained in the SEM with an EDAX GENESIS XM 4i energy-dispersive X-ray spectrometer (EDX, Mahwah, NJ, USA). The analytical distance used for X-ray measurement was 10 mm, which corresponded to a take-off angle of 35°. The measurement was performed at a pressure of 3×10^{-4} Pa (water vapor) and an accelerating voltage of 15 kV. A 0.15×0.13 mm^2 area was scanned and an average spectrum was obtained on the whole area. The elemental composition is calculated from this spectrum, assuming the sample is only composed of aluminum, oxygen, and carbon. This assumption is made after observing the whole EDX spectra and identifying the peaks exhibiting a significant height.

2.8. Surface Topography

Surface profile was obtained for the ablated Al by means of a P17 (KLA Tencor, Milpitas, CA, USA) mechanical profilometer equipment with a scanning load of 0.5 mg. Measurements were performed by using acquisition rate of 50 Hz, a scanning speed of 20 μm/s, and a scanning length of 2 mm, resulting in a scanning time of 100 s and 5000 measured points. Six regions of 1 cm^2 area were ablated on an Al sample. Four measurements were performed on each region; two along the axis of applied pulling forces during shear testing, and two perpendicular to it. The roughness profile was calculated with a cut-off length of 0.25 mm. Roughness parameters Rq (average quadratic height or "root-mean-square" roughness) and Rdq (average quadratic slope) were calculated following ISO 4287 [26], and their average value is reported.

Similar measurements performed on PA sample show that its average quadratic height (Rq) is close to 40 nm. Raw or ablated aluminum exhibit Rq values close to 400 nm or in the range of 0.9–6 μm, respectively. It shows that the PA is very smooth compared to ablated aluminum and that the schematic drawing in Figure 3b is a reasonable representation of "real life" conditions.

2.9. Laser Flash Analysis (LFA)

In order to prepare ablated samples for LFA, they were cut to 1 cm^2 squared geometry, using an Accutom 50 dicing tool from Struers (Ballerup, Denmark). Then, ablated samples were arranged in layered configuration, together with a 1 mm thick polished aluminum sample with the same dimensions as the ablated ones, as shown in Figure 4. Samples were coated on both external faces using Graphit 33 spray from Kontakt Chemie (Iffezheim, Germany) containing 1–5 *w/w* % of graphite powder in order to have a consistent absorbance to the laser beam and consistent emissivity to the IR detector.

Laser flash analysis (LFA) test was performed by a Netzsch LFA 457 Microflash machine (Selb, Germany) at room temperature. A single flash (0.5 ms) from Nd-Yag laser was irradiated on the untreated surface of the ablated aluminum sample, as illustrated in Figure 4. The LFA chamber was filled with dry argon gas during the experimentation in order to reduce the influence of moisture on the measurement of the thermal properties. An infrared detector (InSb photodiode) was used to monitor the temperature transient at the back face. The output voltage of the laser was fixed at 1922 V for all experiments. A duration of 60 ms was used for the acquisition time of the IR detector.

Layered configuration (see Figure 4) was used in order to calculate the TCR generated by the particular geometry of the rough ablated surface in contact with a flat surface [27]. It is difficult to calculate the exact TCR between PA and Al by using LFA with such layered configuration, due to experimental challenges resulting from the low thermal conductance of PA. Therefore, the rough ablated aluminum is brought into contact with another flat aluminum sample. Thermal diffusivity of the layered system is evaluated from the measurement of temperature increase as a function of time at the back face of the polished aluminum, i.e., the side facing the IR sensor. Then, the density, thickness, thermal diffusivity, and specific heat of both materials were predefined to that of pure aluminum, kept constant for all tested samples, and used to calculate the TCR based on the algorithm developed by Hartmann et al. [28].

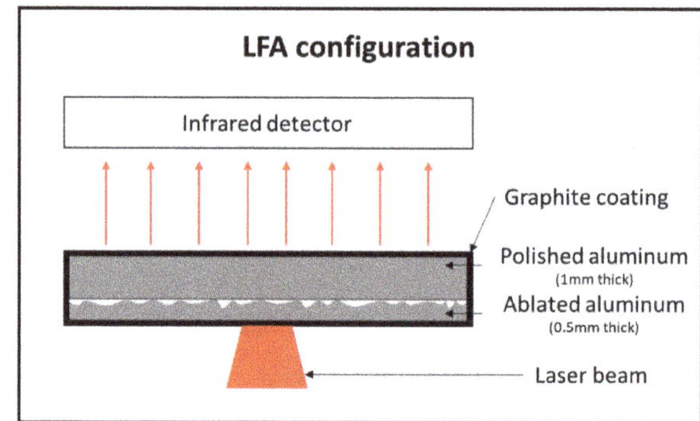

Figure 4. Schematic drawing of laser flash analysis layered configuration.

3. Results

3.1. Joint Strength

The joint area was identified at the joint interface after failure as the area of residues/damages on both materials. The joint area measured in the case of ethanol-wiped aluminum (Al ref) showed 58% of the joint area measured on the corresponding PA indicating a mixture between adhesive and cohesive failure modes. Figure 5 plots the joint area versus shear load at failure for samples that had their aluminum ablated prior to the joining process with the six parameters discussed in this article. Although shear load at failure of the tested samples varies from 524 to 1632 N, linear relation between joint area and shear load indicates constant strength across all samples. A descriptive model was generated to describe the relation between all data points. High coefficient of determination (R^2) of 0.98 illustrates very low variability in the calculated strength, as indicated by a slope of 34.69 MPa. In addition, results show equal and matching joint area on corresponding Al and PA samples, demonstrating prominence of cohesive failure mode for the ablated aluminum and indicating that the reported variation in joint quality is less likely to be a result of variations in interfacial chemical-bonding behavior. Results confirm that laser-ablation parameters have a significant influence on the joint quality, manifested in the joint area as shown in Figure 6, but no influence on the joint strength. Table 2 shows the effect of laser-ablation parameters on joint resistance to shear load.

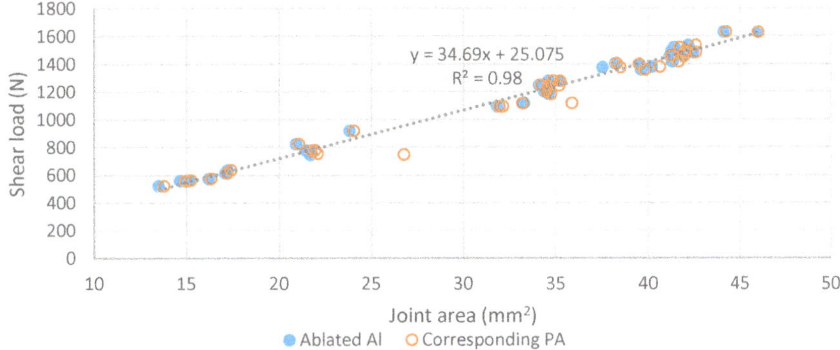

Figure 5. Evolution of joint area, along with increasing joint's resistance to failure.

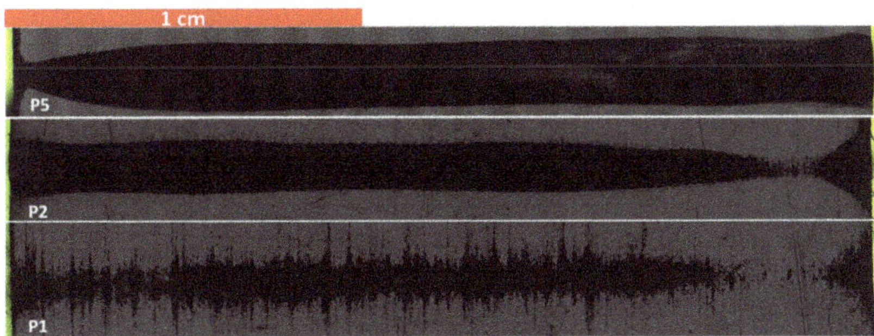

Figure 6. Stitched microscopic images of PA joint area after failure (dark areas), illustrating effect of laser-ablation parameters on joint quality.

Table 2. Effect of laser-ablation parameters on joint resistance to failure.

Ablation Parameters	P1	P2	P3	P4	P5	P5-95%
Average shear load (N)	580 ± 41	800 ± 65	1222 ± 143	1341 ± 172	1415 ± 113	1465 ± 65

3.2. X-Ray Photoelectron Spectroscopy (XPS)

As far as the general composition of the treated aluminum surfaces (see Figure 7) is concerned, it is first observed that laser ablation removes a large part of the adsorbed carbon on the surface, as well as other impurities, resulting in a "cleaning" effect. Concerning the sample which had its surface wiped only with ethanol (Al ref), a high carbon content is visible (29.5 at.%) and a relatively low oxygen concentration (41.1 at.%). After laser pretreatment, a significant decrease in the carbon content on the surface could be noticed, along with an increase in oxygen content. It corresponds to the effects which are generally expected for laser ablation of aluminum: the high energy density of the laser beam has two possible effects, either acting separately or in combination. First, laser ablation certainly causes the removal of surface contaminants, thereby lowering the "masking" effect of the aluminum surface by the contamination top layer. Second, it might contribute to the regeneration of a thicker oxide layer on the Al surface due to high surface temperature during the ablation process. While a minor nitrogen content of 0.7, 0.7, and 0.5 at.% was detected on P1, P2, and P3, samples ablated with parameters P4, P5, and P5-95% illustrated the presence of only aluminum, oxygen, and carbon elements.

Surface chemistry of the different laser-ablated samples do not significantly differ in aluminum, oxygen, and carbon composition. Since no metallic aluminum at 72.7 eV can be found in the high-resolution Al 2p spectra of the ablated samples, Strohmeier's method [29] cannot be used to estimate the oxide thickness. The oxide of the reference samples thickness has been estimated at 5.5 nm by using this method. This implies that the oxide layer is much thicker than the depth limit probed in XPS, which is evaluated in its wide range at 20 nm [30]. The details of the high-resolution spectra of the different elements (C 1*s*, O 1*s*, and Al 2*p*) were very similar across the ablated surfaces, meaning that the surface chemistry is almost identical regardless of the laser-ablation parameters and the resulted joint's strength.

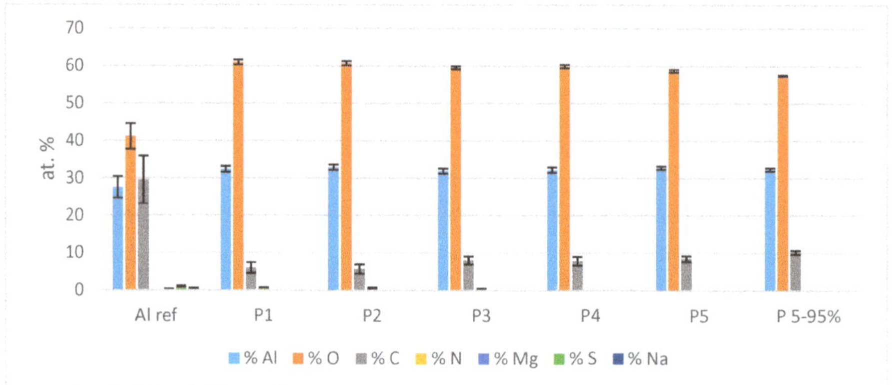

Figure 7. XPS elemental composition of aluminum surfaces.

3.3. Scanning Electron Microscope (SEM)

SEM results shown in Figure 8 illustrates the morphology of the aluminum oxide resulted from the laser-ablation process. The red square shown in 1000× magnification illustrates the area where EDX analysis was performed. Samples ablated with parameter P1, which demonstrated the lowest joint quality, shows relatively high peaks and deep valleys with relatively coarse structures. However, as the joint quality increases, ablated aluminum surfaces are shown to exhibit a smoother surface with finer peak structures, as can be clearly seen on sample P5-95, which resulted in the highest average resistance to failure. Since the failure occurs close to the Al–PA interface (typically 10–15 μm in depth in the PA), the influence of mechanical interlocking cannot be excluded a priori. However, it would be expected that a rougher surface is responsible for a greater mechanical interlocking, thereby reinforcing the joint. The opposite effect is observed here, i.e., the samples exhibiting higher roughness show lower joint quality. It is therefore assumed that mechanical interlocking does not govern the difference of joint resistance to failure.

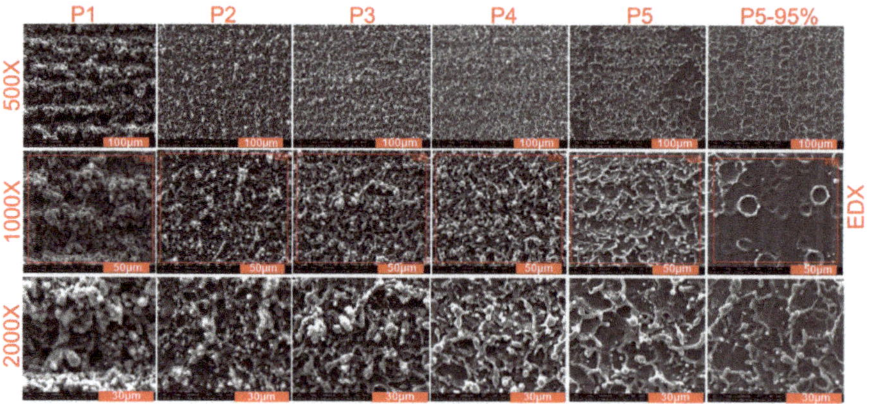

Figure 8. SEM images of ablated surfaces.

3.4. Energy-Dispersive X-Ray Spectroscopy (EDX)

EDX results shown in Figure 9 illustrate atomic percentage of oxygen, carbon, and aluminum. While XPS analyzes a depth which is smaller than 20 nm [30], EDX depth of analysis with the current EDX parameters in use and in case of aluminum oxide would yield a depth close to 2 μm. The carbon element cannot be correctly quantified by the EDX technique and will not be considered in this study.

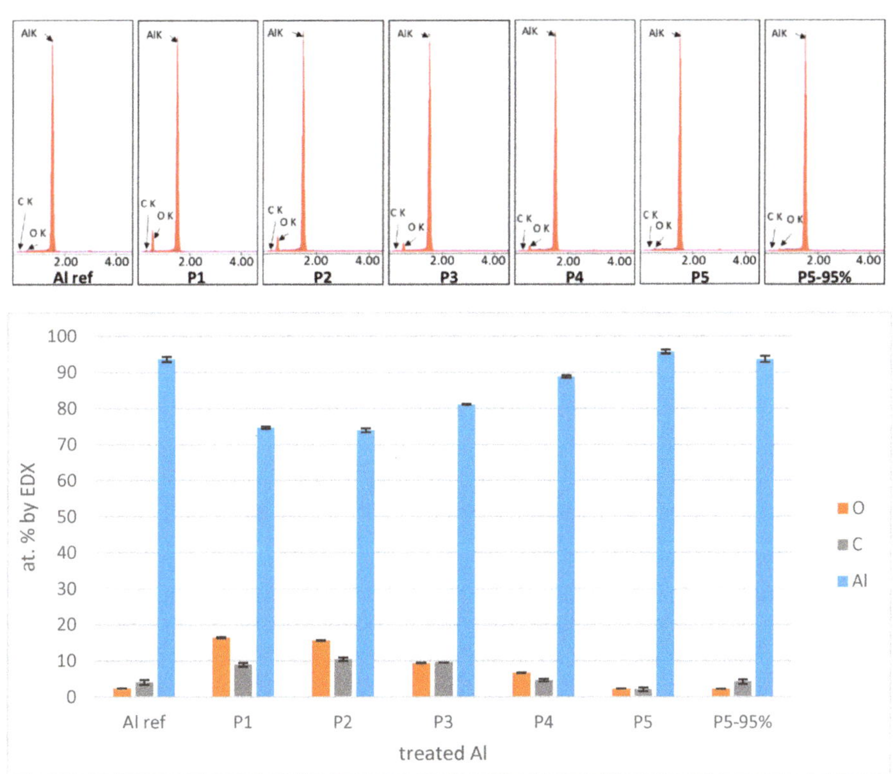

Figure 9. EDX elemental composition of aluminum surfaces.

Interestingly, oxygen concentration detected by EDX increases for the ablation condition. More precisely, this increase is very significant for ablation conditions P1 and P2 (16.4% and 15.6%, respectively), and it is significant to a lesser extent for P3 and P4 (9.4% and 6.7%, respectively), compared to untreated aluminum (2.4%). This leads to the assumption that oxidation of aluminum occurs for ablation conditions P1 to P4 and is particularly large for conditions P1 and P2. This is consistent with XPS results that lead to the conclusion that oxidation occurs at a larger depth than 20 nm for conditions P1 to P4.

A decline in the oxygen peak intensity along the ablated samples from P1 to P5 suggests a decline in aluminum oxide and/or hydroxide layer thickness along the ablation conditions (P1 to P5). EDX is not conclusive for conditions P5 and P5-95%, probably because it is not precise enough at these low concentrations of oxygen (close to 2.4%).

3.5. Surface Topography

The reference aluminum (cleaned with ethanol) exhibited relatively low surface roughness ($Rq = 0.43 \pm 0.26$ μm). Regarding the laser-ablated Al, several topography parameters' values were investigated for correlations with the achieved joint quality. However, no strong correlation was found between Rdq, Rq, and the joint's resistance to failure indicated by the Pearson correlation coefficients of −0.77 and −0.83, respectively. Figure 10 shows the relation between roughness profile parameters ratio Rq/Rdq involved in the TCR model (Equation (1)) with the corresponding joint's resistance to failure, where σ reflects Rq (μm) and m reflects Rdq measured on ablated aluminum surfaces. The correlation between ratio Rq/Rdq and joint's resistance indicated by the Pearson correlation coefficient is −0.93.

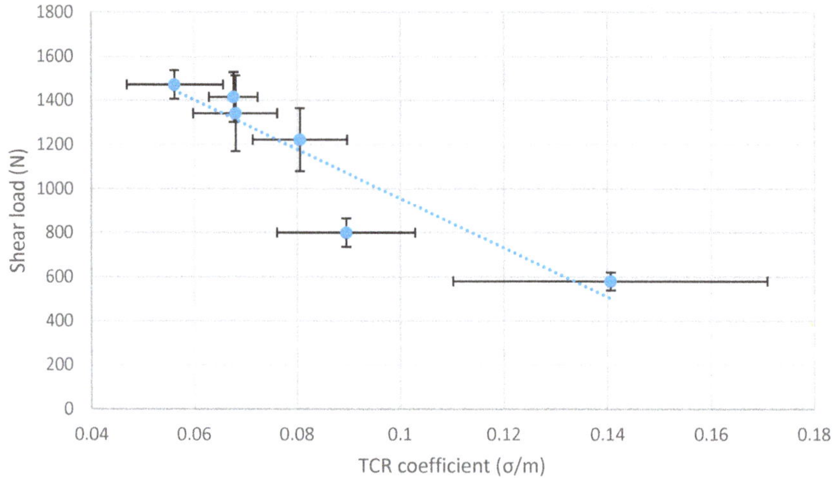

Figure 10. Relation between surface topography and joint's resistance to failure.

3.6. Laser Flash Analysis

In order to fully understand the combined effects of laser-ablation parameters on the thermal transfer through a rough Al surface, LFA layered tests were conducted. Results in Figure 11 show that a decreased TCR across the interface of the layered setup correlates significantly with an increase in the joint's resistance to failure, with a Pearson correlation coefficient of $r = -0.94$. Results confirm that the improvement in joint quality, which is manifested by enhanced joint area and increased resistance to failure, is very likely to be a result of reduced TCR across the interface of the joining partners during the welding process.

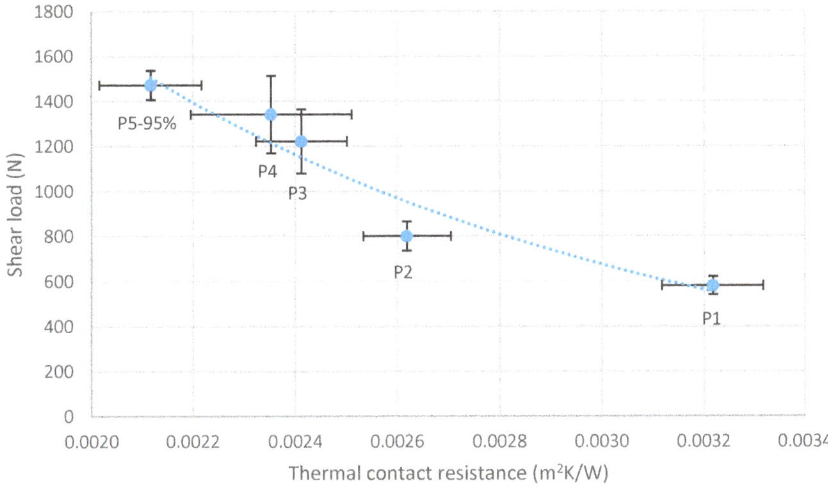

Figure 11. Relation between thermal contact resistance and joint's resistance to failure.

4. Discussion

The lower mechanical resistance of the joint for untreated (reference, only ethanol cleaning) aluminum compared to laser-ablated samples is easily linked to the different surface chemistry. More precisely, laser ablation is responsible to a lower level of carbon contamination and the formation of a relatively thick oxide layer on aluminum. This certainly creates a strong interaction with molten PA during welding, resulting in cohesive failure mode. This is not the case for reference aluminum, which is responsible for the observation of areas of adhesive failure at the interface, thereby reducing the joint resistance.

The difference in joint resistance between laser-ablation conditions clearly comes from a difference in joint area, as stated in 3.1. However, this difference in joint area is difficult to explain. When surface chemistry (as evaluated by XPS) is considered, the only correlation that can be observed is a slight increase of carbon content and a slight decrease of oxygen content from P1 to P5-95, i.e., when shear load increases. An increase in carbon content at the aluminum surface is expected to weaken the interactions at the interface between aluminum and PA and makes the joint weaker. This is contradictory to what is actually observed, in particular if it is also mentioned that failure for ablated samples occurs in PA (cohesive failure), i.e., adhesion at the interface is certainly not responsible for the failure of the assembly.

Interestingly, two good correlations between ablated aluminum properties and shear load are observed. The first one is the decrease of the oxygen concentration, as evaluated by EDX, when shear load increases, and the second is the decrease of Rq/Rdq ratio when shear load increases.

The inverse correlation (Pearson coefficient $r = -0.98$) between EDX oxygen concentration and shear load could probably be explained by a different thermal behavior of the aluminum sample as a function of its content of oxide or hydroxide relatively to metal aluminum. For instance, if it is hypothesized that aluminum oxide is more stable than hydroxide and that it is formed as a layer close to the surface of aluminum, then the aluminum sample can be considered as a two-layer system: metal aluminum in the bulk and aluminum oxide layer at the surface. Since the thermal diffusivity of aluminum oxide (12 mm²/s) is lower than pure aluminum (94 mm²/s) at 300 K [31], this means that a thicker aluminum oxide layer would act as a larger thermal insulator for the transmission of heat to PA during welding. Thus, an increase in oxygen concentration would result in a decrease in volume of molten polymer during welding, leading to a smaller joint area and a decline in the joint's resistance to failure.

Otherwise, the inverse correlation between the *Rq/Rdq* ratio and shear load (Pearson's correlation $r = -0.93$) has to be interpreted in the light of Equation (1), which links the thermal contact resistance and the σ/m ratio as a surface topography characteristic. In terms of phenomena, it means that, when thermal conduction across a "rough interface" (see Figure 3b) is governed by the area of microcontacts, the thermal contact resistance (TCR) increases as the area of microcontacts decreases. A lower value of *Rq/Rdq* ratio leads to a lower value of TCR, meaning that a greater heat flow is conducted across the interface. This leads to the melting of a greater volume of PA during welding, which results in a larger joint and a larger shear load. Furthermore, the fact that TCR evaluated by LFA experiments following Hartmann's method is also well correlated to shear load makes these arguments stronger.

Both phenomena (thermal insulation by aluminum oxide and change of TCR by means of different topography) convincingly explain the increase in shear load from P1 to P5-95. Nevertheless, the authors are currently not able to quantify which phenomenon has the larger effect on this increase, or if one of the two phenomena shall be neglected. Further investigations shall be necessary to answer this question.

Moreover, the correlation between topography and shear load explains why, a posteriori, no correlation was found between the *Ra* topography value (which is strongly correlated to *Rq* value) and shear load in our former article [5]. In such a case, the critical characteristic of the surface topography is its morphology of peaks, evaluated by *Rq/Rdq* ratio, and not its amplitude, which is evaluated by *Ra* or *Rq*.

5. Conclusions

The link between the aluminum surface characteristics after laser ablation and the shear resistance of dissimilar joints formed by laser welding of ablated aluminum with polyamide was investigated. It is observed that shear load at failure of the joint depends on the joint area demonstrating constant joint strength regardless of the ablation parameters. Good correlations between surface characteristics and shear load lead us to make two possible assumptions for explaining the reduction in joint area.

First, it is observed that laser ablation leads to a significant increase of oxygen content in the first micrometers (in depth) of aluminum surface. This might be responsible for a reduction of thermal conduction in the oxygen-rich volume (formation of aluminum oxide/hydroxide), which in turn reduces the quantity of heat transmitted to melt PA during welding.

Second, it is observed that the different topography characteristics obtained after laser ablation of aluminum exhibit different morphology of peaks, which might in turn influence the thermal contact resistance (TCR) between the rough aluminum and the flat PA during welding.

Experimental evaluation of TCR by LFA confirms that thermal-transfer phenomena are probably responsible of changes in shear-load resistance.

This article shed light on the prominence of interfacial thermal-transfer phenomena in the quality of joints obtained by laser welding of a rough ablated aluminum with a polymer.

Author Contributions: Conceptualization, A.A.-S. and J.B.; methodology, A.A.-S., J.B., and P.H.; formal analysis, A.A.-S., J.B., and P.H.; investigation, A.A.-S., J.B., P.H., and R.V.; writing—original draft preparation, A.A.-S.; writing—review and editing, J.B., P.H., R.V., L.H., and P.P.; supervision, L.H. and P.P.; project administration, P.P.; funding acquisition, P.P., J.B., and L.H.

Funding: This research was funded by FNR (Luxembourg) and DGO6 (Belgium) through M-era.net, under project LaserSTAMP.

Acknowledgments: Jean-Luc Biagi (LIST) is gratefully acknowledged for his skillful characterization of aluminum surfaces by SEM. The authors would like to thank Sébastien Depaifve (LIST) and Daniel Schmidt (LIST) for the interesting insights they provided about thermal interface materials and thermal transfer across interfaces.

Conflicts of Interest: The authors declare no conflicts of interest. The funders had no role in the design of the study; in the collection, analyses, or interpretation of data; in the writing of the manuscript, or in the decision to publish the results.

References

1. Lamberti, C.; Solchenbach, T.; Plapper, P.; Possart, W. Laser assisted joining of hybrid polyamide-aluminum structures. *Phys. Procedia* **2014**, *56*, 845–853. [CrossRef]
2. Katayama, S.; Kawahito, Y. Laser direct joining of metal and plastic. *Scr. Mater.* **2008**, *59*, 1247–1250. [CrossRef]
3. Wahba, M.; Kawahito, Y.; Katayama, S. Laser direct joining of AZ91D thixomolded Mg alloy and amorphous polyethylene terephthalate. *J. Mater. Process. Technol.* **2011**, *211*, 1166–1174. [CrossRef]
4. Arai, S.; Kawahito, Y.; Katayama, S. Effect of surface modification on laser direct joining of cyclic olefin polymer and stainless steel. *Mater. Des.* **2014**, *59*, 448–453. [CrossRef]
5. Al-Sayyad, A.; Bardon, J.; Hirchenhahn, P.; Santos, K.; Houssiau, L.; Plapper, P. Aluminum pretreatment by a laser ablation process: Influence of processing parameters on the joint strength of laser welded aluminum—Polyamide assemblies. *Procedia CIRP* **2018**, *74*, 495–499. [CrossRef]
6. Bergmann, J.P.; Stambke, M. Potential of laser-manufactured polymer-metal hybrid joints. *Phys. Procedia* **2012**, *39*, 84–91. [CrossRef]
7. Zhang, Z.; Shan, J.G.; Tan, X.H.; Zhang, J. Effect of anodizing pretreatment on laser joining CFRP to aluminum alloy A6061. *Int. J. Adhes. Adhes.* **2016**, *70*, 142–151. [CrossRef]
8. Klotzbach, A.; Langer, M.; Pautzsch, R.; Standfuß, J.; Beyer, E. Thermal direct joining of metal to fiber reinforced thermoplastic components. *J. Laser Appl.* **2017**, *29*, 022421. [CrossRef]
9. Amend, P.; Pfindel, S.; Schmidt, M. Thermal joining of thermoplastic metal hybrids by means of mono- and polychromatic radiation. *Phys. Procedia* **2013**, *41*, 98–105. [CrossRef]
10. Holtkamp, J.; Roesner, A.; Gillner, A. Advances in hybrid laser joining. *Int. J. Adv. Manuf. Technol.* **2010**, *47*, 923–930. [CrossRef]
11. Georgiev, G.L.; Baird, R.J.; McCullen, E.F.; Newaz, G.; Auner, G.; Patwa, R.; Herfurth, H. Chemical bond formation during laser bonding of Teflon® FEP and titanium. *Appl. Surf. Sci.* **2009**, *255*, 7078–7083. [CrossRef]
12. Roesner, A.; Scheik, S.; Olowinsky, A.; Gillner, A.; Reisgen, U.; Schleser, M. Laser assisted joining of plastic metal hybrids. *Phys. Procedia* **2011**, *12*, 373–380. [CrossRef]
13. Cui, T.; Li, Q.; Xuan, Y. Characterization and application of engineered regular rough surfaces in thermal contact resistance. *Appl. Therm. Eng.* **2014**, *71*, 400–409. [CrossRef]
14. Zhang, P.; Cui, T.; Li, Q. Effect of surface roughness on thermal contact resistance of aluminium alloy. *Appl. Therm. Eng.* **2017**, *121*, 992–998. [CrossRef]
15. Prasher, R. Thermal interface materials: Historical perspective, status, and future directions. *Proc. IEEE* **2006**, *94*, 1571–1586. [CrossRef]
16. Yovanovich, M.M. Four decades of research on thermal contact, gap, and joint resistance in microelectronics. *IEEE Trans. Compon. Packag. Technol.* **2005**, *28*, 182–206. [CrossRef]
17. Greenwood, J.A.; Williamson, J.B.P. Contact of nominally flat surfaces. *Proc. R. Soc. London Ser. A Math. Phys. Sci.* **1966**, *295*, 300–319. [CrossRef]
18. Greenwood, J.A. The area of contact between rough surfaces and flats. *J. Lubr. Technol.* **1967**, *89*, 81–87. [CrossRef]
19. Greenwood, J.A.; Tripp, J.H. The contact of two nominally flat rough surfaces. *Proc. Inst. Mech. Eng.* **1970**, *185*, 625–633. [CrossRef]
20. Cooper, M.G.; Mikic, B.B.; Yovanovich, M.M. Thermal contact conductance. *Int. J. Heat Mass Transf.* **1969**, *12*, 279–300. [CrossRef]
21. Mikić, B.B. Thermal contact conductance; theoretical considerations. *Int. J. Heat Mass Transf.* **1974**, *17*, 205–214. [CrossRef]
22. Sayles, R.S.; Thomas, T.R. Thermal conductance of a rough elastic contact. *Appl. Energy* **1976**, *2*, 249–267. [CrossRef]
23. Yovanovich, M.M.; Marotta, E.E. Thermal spreading and contact resistances. In *Heat Transfer Handbook*; Bejan, A., Kraus, A.D., Eds.; Wiley: Hoboken, NJ, USA, 2003; pp. 261–395.
24. Bahrami, M.; Yovanovich, M.M.; Marotta, E.E. Thermal joint resistance of polymer-metal rough interfaces. *J. Electron. Packag.* **2006**, *128*, 23. [CrossRef]

25. Al-Sayyad, A.; Bardon, J.; Hirchenhahn, P.; Mertz, G.; Haouari, C.; Houssiau, L.; Plapper, P. Influence of laser ablation and plasma surface treatment on the joint strength of laser welded aluminum-polyamide assemblies. In Proceedings of the JNPLI 2017, Strasbourg, France, 13–14 September 2017.
26. *ISO 4287–Geometrical Product Specifications (GPS)—Surface Texture: Profile Method—Terms, Definitions and Surface Texture Parameters*; International Organization for Standardization: Geneva, Switzerland, 1997.
27. Corbin, S.F.; Turriff, D.M. Thermal diffusivity by the laser flash technique. *Charact. Mater.* **2002**. [CrossRef]
28. Hartmann, J.; Nilsson, O.; Fricke, J. Thermal diffusivity measurements on two-layered and three-layered systems with the laser-flash method. *High Temp. Press.* **1993**, *25*, 403.
29. Strohmeier, B.R. An ESCA method for determining the oxide thickness on aluminum alloys. *Surf. Interface Anal.* **1990**, *15*, 51–56. [CrossRef]
30. Ratner, B.D.; Castner, D.G. Electron spectroscopy for chemical analysis. In *Surface Analysis—The Principal Techniques*; John Wiley & Sons, Ltd.: Chichester, UK, 2009; pp. 47–112.
31. King, J.A. *Materials Handbook for Hybrid Microelectronics*; Artech House: Boston, MA, USA, 1988.

© 2019 by the authors. Licensee MDPI, Basel, Switzerland. This article is an open access article distributed under the terms and conditions of the Creative Commons Attribution (CC BY) license (http://creativecommons.org/licenses/by/4.0/).

 *coatings* 

Article

Thermoelastic Response Induced by Volumetric Absorption of Uniform Laser Radiation in a Half-Space

Ismail M. Tayel

Department of Mathematics, College of Science Al-Zulfi, Majmaah University, Majmaah 11952, Saudi Arabia; i.tayel@mu.edu.sa

Received: 4 February 2020; Accepted: 22 February 2020; Published: 2 March 2020

Abstract: In this paper, the thermoelastic response in a generalized thermoelastic half-space induced by absorption of a penetrating pulsed laser radiation inside the medium is studied using the generalized theory with Dual-Phase-Lag (DPL). The surface of the target is considered stresses free and exposed to temperature-dependent heat losses. Laplace integral transform is used analytically for obtaining the general solution, while its inverse is carried out numerically. The copper element is used as an application to compare the predictions induced by volumetric absorption of the Dual-Phase-lag theory with those for Lord–Shulman (LS) and classical coupled (CTE) theories, moreover the response induced by volumetric absorption for (LS) and (CTE) models in this work were compared with those induced by surface absorption in a previous work.

Keywords: generalized thermoelasticity; laser radiation; volumetric absorption; thermal stresses; cooling effect

---

1. Introduction

A rapid growth in scientific research of the thermoelasticity field was observed after Biot (1956) [1] introduced the coupled theory of thermoelasticity (CTE) to fade the flaws of the classical uncoupled theory, stating that the elastic disturbance has no influence on temperature and the temperature waves travel with infinite speeds. Although this theory repairs the first defect by coupling between temperature and strain, the energy equation of this theory was based on Fourier's law and, thus, its heat waves still travel with infinite speed. Employing Biot's model Hetnarski [2,3] has discussed some remarkable problems in this direction. Furthermore, the work done by Boley and Tolins [4] considered a good contribution.

The defect of the infinite speed in Biot's theorem was eliminated by the generalization introduced by Lord and Shulman in 1967, this generalization was known as the generalized theory of thermoelasticity with one relaxation time (LS) [5]. The energy equation of the latter theory was constructed on a new law of heat conduction instead of Fourier's law and, thus, the defect of infinite speed was repaired. As significant contributions in the context of this theory, we mention the proofs of uniqueness theorems under different conditions by Ignaczak [6,7]. Furthermore, Sherief el al. [8–10] obtained solutions of some thermoelastic problems by employing Lord and Shulman's theory.

Many models such as [11–13] have been introduced to be generalizations to Biot's theory, in which the generated heat waves travel with finite speeds. One of the substantial models was the generalized theory with dual-phase-lag (DPL), which was developed by Ozisik and Tzou [14] and Tzou [15,16]. In this model, Fourier's law was replaced by an approximation that includes two distinct time translations signifying the phase lags of temperature gradient and heat flux. RoyChoudhuri [17] employed the DPL model to study the disturbances in an elastic 1D half-space subjected to two different boundary conditions. Using the DPL model, Aboelregal [18] obtained the solution of a 1D semi-infinite medium whose boundary plane was exposed to thermal shock. EL-Karamany and

Ezzat [19] published a considerable study on the Dual-Phase-Lag model. Abbas and Zenkour [20] studied the dual-phase-lag model on thermoelastic interactions in a half-space exposed to a ramp-type heating by applying the finite element method. Marin et al. [21] studied the well-posed dual-phase-lag model of a thermoelastic dipolar material.

One of the significant sources leading to thermal influences is the pulsed laser, which gives a high density in small periods of time and this, in turn, yields a great importance economically. Since lasers were founded, many applications have been made in science, mechanics and engineering based on their specific properties [22]. Many theoretical studies have been conducted to study thermoelastic responses induced by laser radiation [23–25]. Henain et al. [26] studied the surface illumination of a 2D thermoelastic homogeneous semi-infinite medium. Yossef and EL-Bary [27] applied four thermoelastic theories to study the response in an elastic half-space induced by laser pulse whose temporal profile is non-Gaussian. Allam and Tayel [28] obtained a solution for a 1D thermoelastic functionally graded half-space heated uniformly by a surface absorption of pulsed laser having Gaussian temporal profile. Abbas and Marin [29] studied the thermoelastic interaction induced by laser pulse in a 2D half-space and introduced an analytical solution in the context of the generalized theory with one relaxation time. Tayel and Hassan [30] employed the fractional order theory of thermoelasticity to study the effect of the fractional parameter in the existence and absence of heat losses in a thermoelastic half-space heated uniformly by surface absorption of laser radiation.

The aims of this work are to investigate the thermoelastic interactions in a half-space induced by volumetric absorption of laser radiation in the existence and absence of heat losses by employing three different theories of thermoelasticity, namely, DPL, LS, and CTE, and to study the influences of the phase lag of the temperature gradient of the dual-phase-lag theory. Moreover, the response of LS and CTE models will be compared with the response induced by surface absorption technique in a previously published paper [30].

2. Problem Formulation and Basic Equation

We consider a laser beam to be incident uniformly in z-direction perpendicular to a homogeneous and isotropic thermoelastic half-space ($z \geqslant 0$) whose initial temperature is T_0. The irradiated surface ($z = 0$) is taken to be traction free and subjected to heat losses whose coefficient is temperature-dependent. At the irradiated surface, a part of the laser radiation will be reflected, while the rest will be absorbed into the medium to be converted into heat and, consequently, thermoelastic waves will be generated.

Due to the uniform illumination of the surface ($z = 0$), the problem will be 1D and all studied fields will be functions of z and t only, and consequently, the displacement vector **u** will have one component, namely $w(z, t)$, while the other components are vanishing.

Following the Dual-Phase-Lag theory (DPL) [15,16], the equations governing the thermoelastic waves in the absence of body forces for a one-dimensional problem are

- The heat conduction equation:

$$k\left(1 + \tau_1 \frac{\partial}{\partial t}\right)\frac{\partial^2 T}{\partial z^2} = \left(1 + \tau_2 \frac{\partial}{\partial t}\right)\left(\rho C_E \frac{\partial T}{\partial t} + T_0 \gamma \frac{\partial^2 w}{\partial t \partial z} - Q\right). \tag{1}$$

where T, k, ρ, c_E, Q, T_0 stand for the absolute temperature, thermal conductivity, denisty, specific heat at costant strain, heat source per unit volume, and reference temperature, which chosen such that $|(T - T_0)/T_0| \ll 1$. Furthermore, $\gamma = (3\lambda + 2\mu)\alpha_T$, in which λ and μ are Lamé's constant and α_T is the coefficient of linear thermal expansion. τ_1 and τ_2 are the phase lag of the temperature gradient and heat flux, respectively, where $(0 \leq \tau_1 < \tau_2)$.

The energy equations of LS and CTE models can be generated from Equation (1) by setting $\tau_1 = 0$ and $\tau_1 = \tau_2 = 0$, respectively.

- According to the volumetric absorption technique, the heat source, $Q(z, t)$ takes the form:

$$Q(z, t) = A_0 q_0 g(t)\xi e^{-\xi z}. \tag{2}$$

where A_0 is the transition coefficient of the irradiated surface, ξ is the linear absorption coefficient of the material, q_0 is the maximum value of the laser power density and $g(t)$ is the time dependent laser pulse profile.

Using Equation (2), Equation (1) can be written as

$$k\left(1 + \tau_1 \frac{\partial}{\partial t}\right)\frac{\partial^2 T}{\partial z^2} = \left(1 + \tau_2 \frac{\partial}{\partial t}\right)\left(\rho C_E \frac{\partial T}{\partial t} + T_0 \gamma \frac{\partial^2 w}{\partial t \partial z} - A_0 q_0 g(t)\xi e^{-\xi z}\right).$$
(3)

- The equation of motion:

$$\frac{\partial \sigma_{zz}}{\partial z} = \rho \frac{\partial^2 w(z,t)}{\partial t^2}.$$
(4)

where σ_{zz} is the normal stress.

For a 1D problem, the strain components are

$$e_{zz} = \frac{\partial w}{\partial z}, \quad e_{xx} = e_{yy} = e_{zx} = e_{zy} = e_{xy} = 0.$$
(5)

Thus, the volume dilatation will be

$$e = e_{zz} + e_{xx} + e_{yy} = \frac{\partial w}{\partial z}.$$
(6)

The non-vanishing stresses are :

$$\sigma_{zz} = 2\mu \frac{\partial w}{\partial z} + \lambda e - \gamma\theta,$$
(7a)

$$\sigma_{xx} = \lambda \frac{\partial w}{\partial z} - \gamma\theta,$$
(7b)

$$\sigma_{yy} = \sigma_{xx}.$$
(7c)

where $\theta = T - T_0$ is the temperature increment.

Using Equation (7a), Equation (4) can be written as

$$(\lambda + 2\mu)\frac{\partial^2 w}{\partial z^2} - \gamma\frac{\partial \theta}{\partial z} = \rho\frac{\partial^2 w}{\partial t^2}.$$
(8)

Now, the boundary conditions can be expressed as:

$$k\left(1 + \tau_1 \frac{\partial}{\partial t}\right)\frac{\partial \theta}{\partial z}\bigg|_{z=0} = \left(1 + \tau_2 \frac{\partial}{\partial t}\right)h\theta(0,t),$$
(9a)

$$\sigma_{zz}\bigg|_{z=0} = 0,$$
(9b)

$$\theta, w, \sigma_{zz} \to 0 \quad as \quad z \to \infty.$$
(9c)

where $\theta(0,t)$ is the surface temperature and h is the heat transfer coefficient.

Also, the initial conditions are:

$$\left.\begin{array}{l}
\theta(z,t)\bigg|_{t=0} = \dfrac{\partial \theta(z,t)}{\partial t}\bigg|_{t=0} = 0, \quad w(z,t)\bigg|_{t=0} = \dfrac{\partial w(z,t)}{\partial t}\bigg|_{t=0} = 0 \\[2ex]
\sigma_{ij}(z,t)\bigg|_{t=0} = \dfrac{\partial \sigma_{ij}(z,t)}{\partial t}\bigg|_{t=0} = 0.
\end{array}\right\}$$
(10)

For simplicity, we introduce the following non-dimensional variables:

$$(z^*, w^*) = (z, w)c\eta \qquad \theta^* = \frac{\gamma\theta}{(\lambda + 2\mu)} \qquad \xi^* = \frac{\xi}{c\eta}$$

$$(t^*, \tau^*_{1,2}) = (t, \tau_{1,2})c^2\eta \qquad q_0^* = \frac{\gamma q_0}{(\lambda + 2\mu)c\rho C_E} \qquad h^* = \frac{h}{c\rho C_E}$$

$$\sigma^*_{ij} = \frac{\sigma_{ij}}{\mu} \qquad c = \sqrt{\frac{\lambda + 2\mu}{\rho}} \qquad \eta = \frac{\rho C_E}{k}.$$

Making use of the non-dimensional variables after dropping the stars, Equations (3), (8) and (7) can be written respectively as:

$$\left(1 + \tau_1 \frac{\partial}{\partial t}\right) \frac{\partial^2 \theta}{\partial z^2} = \left(1 + \tau_2 \frac{\partial}{\partial t}\right) \left(\frac{\partial \theta}{\partial t} + \epsilon \frac{\partial^2 w}{\partial t \partial z} - A_0 q_0 \xi e^{-\xi z} g(t)\right). \tag{11}$$

$$\frac{\partial^2 w}{\partial z^2} - \frac{\partial \theta}{\partial z} = \frac{\partial^2 w}{\partial t^2}. \tag{12}$$

$$\sigma_{zz} = g_1 \left(\frac{\partial w}{\partial z} - \theta\right), \tag{13a}$$

$$\sigma_{xx} = g_2 \frac{\partial w}{\partial z} - g_1 \theta, \tag{13b}$$

$$\sigma_{xx} = \sigma_{yy}. \tag{13c}$$

where

$$\epsilon = \frac{\gamma^2 T_0}{\rho C_E (\lambda + 2\mu)}, \quad g_1 = \frac{\lambda + 2\mu}{\mu}, \quad g_2 = \frac{\lambda}{\mu}.$$

The non-dimensional form of (9) is given as:

$$\left(1 + \tau_1 \frac{\partial}{\partial t}\right) \frac{\partial \theta}{\partial z}\bigg|_{z=0} = \left(1 + \tau_2 \frac{\partial}{\partial t}\right) h\theta(0, t), \tag{14a}$$

$$\sigma_{zz}\bigg|_{z=0} = 0, \tag{14b}$$

$$\theta, w, \sigma_{zz} \to 0 \quad as \quad z \to \infty. \tag{14c}$$

3. Problem Solution

Introducing the Laplace integral transform with respect to the time [31]:

$$\tilde{f}(z, s) = \int_0^\infty f(z, t) e^{-st} dt. \tag{15}$$

where $\tilde{f}(z, s)$ is the Laplace transform of $f(z, t)$ and s is the temporal angular frequency.

Applying (15) on Equations (11) and (12), one gets respectively

$$(D^2 - \Lambda s)\tilde{\theta} - \epsilon s \Lambda D\tilde{w} = -A_0 q_0 \Lambda \xi e^{-\xi z} \tilde{g}(s). \tag{16}$$

$$(D^2 - s^2)\tilde{w} - D\tilde{\theta} = 0. \tag{17}$$

where $\Lambda = \frac{(1+\tau_2 s)}{(1+\tau_1 s)}$ and $\tilde{g}(s)$ is the Laplace transform of $g(t)$.

Applying the operator $(D^2 - s^2)$ on Equation (16) and $\epsilon s \Lambda D$ on Equation (17) then adding, one obtains the following fourth-order nonhomogeneous differential equation

$$(D^4 - (s^2 + s\Lambda(1 + \epsilon))D^2 + s^3\Lambda)\tilde{\theta} = -A_0 q_0 \Lambda \xi(\xi^2 - s^2)e^{-\xi z}\tilde{g}(s). \tag{18}$$

The last equation possesses the following characteristic equation

$$m^4 - b_1(s)m^2 + b_2(s) = 0. \tag{19}$$

where $b_1(s) = s^2 + s\Lambda(1 + \epsilon)$ and $b_2(s) = s^3\Lambda$.

It follows from Equation (18), that

$$(D^2 - m_1^2)(D^2 - m_2^2)\tilde{\theta} = -A_0 q_0 \Lambda \xi(\xi^2 - s^2)e^{-\xi z}\tilde{g}(s). \tag{20}$$

where $m_i^2, (i = 1, 2)$ are the roots of Equation (19), in which:

$$m_i^2 = \frac{s^2 + s\Lambda(1 + \epsilon) \pm \sqrt{(s^2 + s\Lambda(1 + \epsilon))^2 - 4s^3\Lambda}}{2}. \tag{21}$$

with the aid of the undermined coefficients method, the general solution of (20) is found to be

$$\tilde{\theta} = \sum_{i=1}^{2} B_i(s)e^{-m_i z} + H(s)e^{-\xi z}, \tag{22}$$

with:

$$H = \frac{-A_0 q_0 \Lambda \xi(\xi^2 - s^2)\tilde{g}(s)}{(\xi^2 - m_1^2)(\xi^2 - m_2^2)}. \tag{23}$$

Following a similar manner to obtain $\tilde{w}$, it follows

$$(D^2 - m_1^2)(D^2 - m_2^2)\tilde{w} = A_0 q_0 \Lambda \xi^2 e^{-\xi z}\tilde{g}(s). \tag{24}$$

Solving Equation (24), one gets

$$\tilde{w} = \sum_{i=1}^{2} N_i(s)e^{-m_i z} + G(s)e^{-\xi z}, \tag{25}$$

with

$$G(s) = \frac{A_0 q_0 \Lambda \xi^2 \tilde{g}(s)}{(\xi^2 - m_1^2)(\xi^2 - m_2^2)}. \tag{26}$$

Substituting for $\tilde{w}$ and $\tilde{\theta}$ in Equation (17), we obtain:

$$N_i(s) = \frac{-m_i}{m_i^2 - s^2}B_i(s) \quad (i = 1, 2), \tag{27a}$$

$$G(s) = \frac{-\xi}{\xi^2 - s^2}H(s). \tag{27b}$$

Consequently, we can write

$$\tilde{w} = -\sum_{i=1}^{2}\left(\frac{m_i}{m_i^2 - s^2}\right)B_i(s)e^{-m_i z} - \left(\frac{\xi}{\xi^2 - s^2}\right)H(s)e^{-\xi z}. \tag{28}$$

Applying (15) on Equation (13), then substituting for $\tilde{\theta}$ and $\tilde{w}$, one find:

$$\tilde{\sigma}_{zz} = g_1 \left[\left(\frac{m_i^2}{m_i^2 - s^2} - 1 \right) B_i(s) e^{-m_i z} + \left(\frac{\xi^2}{\xi^2 - s^2} - 1 \right) H(s) e^{-\xi z} \right], \tag{29a}$$

$$\tilde{\sigma}_{xx} = \left[\left(g_2 \frac{m_i^2}{m_i^2 - s^2} - g_1 \right) B_i(s) e^{-m_i z} + \left(g_2 \frac{\xi^2}{\xi^2 - s^2} - g_1 \right) H(s) e^{-\xi z} \right], \tag{29b}$$

$$\tilde{\sigma}_{xx} = \tilde{\sigma}_{yy}. \tag{29c}$$

By Applying (15) on (14) and using Equations (22) and (29a), one gets:

$$-\sum_{i=1}^{2} m_i B_i = \xi H(s) + \Lambda h \tilde{\theta}(0,s), \tag{30a}$$

$$\sum_{i=1}^{2} f_i B_i = -\Omega H(s). \tag{30b}$$

with

$$\Omega = \frac{\xi^2}{\xi^2 - s^2} - 1, \quad f_i = \frac{m_i^2}{m_i^2 - s^2} - 1, (i = 1, 2)$$

where $\tilde{\theta}(0,s)$ is the Laplace transform of $\theta(0,t)$.

Solving Equations (30a) and (30b) for $B_i, (i = 1, 2)$, one obtains:

$$B_1 = \frac{(h\Lambda\tilde{\theta}(0,s) + H\xi)f_2 - H\Omega m_2}{-f_2 m_1 + f_1 m_2}, \tag{31a}$$

$$B_2 = \frac{-(h\Lambda\tilde{\theta}(0,s) + H\xi)f_1 + H\Omega m_1}{-f_2 m_1 + f_1 m_2}. \tag{31b}$$

Equations (31) contains $\tilde{\theta}(0,s)$, which still un-known, to complete the solution we shall substitute B_1 and B_2 in (22), one gets after setting $z = 0$, the following equation:

$$\tilde{\theta}(0,s) = \frac{H(f_2(\xi - m_1) + f_1(m_2 - \xi) + \Omega(m_1 - m_2))}{-f_2(h\Lambda + m_1) + f_1(h\Lambda + m_2)}. \tag{32}$$

4. Inverse Laplace Transformation

The formulas in Equations (22), (28) and (29) and the roots in Equation (21) predict great difficulties to be converted into the time domain by the usual analytical methods. So, the idea of utilizing a numerical methods will be acceptable to obtain inverse Laplace transform. Following [16], the method of Riemann-sum approximation, with $\eta t = 4.7$ will be adopted using the following formula:

$$f(t) = \frac{e^{\eta t}}{t} \left[\tfrac{1}{2}\tilde{f}(\eta) + Re \sum_{n=1}^{K} (-1)^n \, \tilde{f}\left(\eta + \frac{in\pi}{t}\right) \right]. \tag{33}$$

where K represents the number of terms, which should be sufficiently large and i is the imaginary unit numper.

5. Application and Computation

To demonstrate the consistency of the obtained results, an illustrative example is given for a copper half-space whose surface is illuminated by a pulsed laser assuming that $\tau_2 = 75 \times 10^{-5}$ and

$\zeta = 10^5 \quad m^{-1}$. In order to compare the results of the present paper with those considered the surface absorption, the optical and physical parameters have to be used as [30]:

$$\rho = 8954 \quad kg/m^3 \qquad \lambda = 7.76 \times 10^{10} \quad kg/(m.s^2) \qquad \mu = 3.86 \times 10^{10} \quad kg/(m.s^2)$$
$$C_E = 383.1 \quad J/(kg.K) \qquad T_0 = 293 \quad K \qquad \alpha_T = 1.78 \times 10^{-5} \quad K^{-1}$$
$$k = 386 \quad W/(m.K) \qquad t_0 = 3 \times 10^{-3} \quad s \qquad \delta = 10^{-3} \quad s$$
$$h = 200 \quad W/(m^2.K) \qquad q_0 = 10^4 \quad W/m^2 \qquad A_0 = 0.01.$$

Consider the time dependent laser pulse profile $g(t)$ to be given by a Gaussian distribution as:

$$g(t) = e^{-\left(\frac{t-t_0}{\delta}\right)^2}. \tag{34}$$

where δ is the moment at which the laser beam intensity is reduced to $\frac{1}{e}$ and t_0 is the moment at which $g(t)$ becomes maximum.

6. Results

Figure 1 represents beside the curve of the chosen pulse profile $g(t)$, the time dependent surface temperature $\theta(0, t)$ for three different theories, namely LS, DPL and CTE calculated for $h = 0$, the DPL curve calculated for $\tau_1 = 30 \times 10^{-5}$. The curves show, $\theta(0, t)$ increases with increasing the exposure time until the temperature reaches its maximum and then begins to decrease. This behavior can be interpreted as follows: at the beginning of the irradiation process, the conversion rate of the absorbed energy into heat is greater than the rate of the conducted one to the surrounding; this leads to piling up the heat energy in the vicinity of the illuminated surface and, thus, temperature rise. The increment of $\theta(0, t)$ is continued until the converted heat energy becomes equivalent to the conducted one; at this moment, the temperature reaches its maximum. After that, the absorbed energy begins to decrease and thus the conduction becomes greater than the absorption leading to decrease the temperature. As seen, the maximum value of the surface temperature is greater in the case of the generalized theories of LS and DPL than the case of the classical coupled theory of CTE. For the three curves, the maximum is shifted to greater times than the maximum of $g(t)$, but the CTE needs more time than the other theories to reach its maximum. The observed behavior can be interpreted as follows: beside the infinite speed of the heat waves of the CTE model, which leads to deeper spread of the temperature inside the medium, two different phenomena affect the spatial and temporal temperature distribution, namely, the heat conductivity described by the gradient of the temperature and the expansion of the material. While the first leads to heating the surroundings, the second will consume the heat energy in expanding the distances between particles and, thus, its potential energy increases. As observed in a later figure the CTE model having the greater spreading and the greater gradient of the spatial temperature close to the irradiated surface, so it has the smallest surface temperature. Since the maximum temperature occurs as the absorbed energy is equal to the consumed one, the LS and DPL maximum must occur more near the maximum of the laser profile $g(t)$ than that of the CTE, which consumes more energy in penetrating more into the medium. According to [30], the maximum values of $\theta(0, z)$ for the models of CTE and LS induced by the surface absorption is greater than the case of the volumetric absorption by approximately 20% and its shift towards t-values is less.

Figure 2 represents $\theta(0, t)$ of the three theories LS, DPL and CTE calculated for $h \neq 0$, the DPL curve calculated for $\tau_1 = 30 \times 10^{-5}$. The figure shows a different behavior from Figure 1, where the three curves almost match with the curve of the pulse profile $g(t)$; this behavior is due to the conductivity of the martial together with the cooling effect. As in Figure 1, the curves of LS and CTE induced by volumetric absorption are smaller than the curves of the surface absorption [30].

Figure 3 represents $\theta(0, t)$ of the DPL model calculated for $h = 0$ and different values of τ_1 at a fixed value of τ_2. The figure indicates that as τ_1 takes value near τ_2, $\theta(0, t)$ behaves approximately like the classical coupled theory (CTE) and as τ_1 takes values far from τ_2, $\theta(0, t)$ behaves approximately as the generalized theories in which this behavior is valid until a certain value approximately $\tau_1 = 1 \times 10^{-5}$ after this value the curves will be coincided for any values of τ_1. The figure shows also that as τ_1 decreases, the maximum of $\theta(0, t)$ increases. After $\theta(0, t)$ reaches its maximum value the slop of the curves are inverted with respect to the slop before the maximum value.

Figure 4 illustrates the spatial distribution of the temperature of the DPL model calculated for $h = 0$ and $\tau_1 = 30 \times 10^{-5}$ at several times. The curves reveal a clear finite velocity which appears through the strong gradient at diverse locations and the increasing deep penetration as the time increase. The figure agrees with Figure 1, where its maximum temperature is located at the irradiated surface and occurs at a time greater than the time of the pulse profile $g(t)$. After the time at which the temperature reaches its maximum ($t = 3.4 \times 10^{-3}$), a small gradient near the irradiated surface is observed owing to the decay of the absorbed energy inside the medium.

Figure 5 illustrates the spatial distribution of the temperature of three different models, LS, DPL and CTE, calculated for $h = 0$ and $t = 4 \times 10^{-3}$, the DPL curve calculated for $\tau_1 = 30 \times 10^{-5}$. The figure shows that the CTE model penetrates into the medium more than the other models (DPL and LS) and has the greatest gradient in the vicinity of the irradiated surface. This behavior is due to its infinite velocity of propagation. As seen, the maximum temperature of the LS model is smaller than the maximum of the other models and occurs near the surface of the target. This behavior is due to its relatively greater displacement at the irradiated surface which can be seen clearly in a later figure. This leads to store the heat energy in a mechanical potential energy and, therefore, slightly cools the surface due to heat conduction. Furthermore, the DPL model appears as a case between the LS and CTE. For the CTE and LS in the case of volumetric absorption, the temperature is smaller and the penetration depth is greater compared with the corresponding case of the surface absorption [30].

Figure 6 illustrates the spatial distribution of the temperature of the three different models (LS, DPL, and CTE) calculated for $h \neq 0$ at $t = 4 \times 10^{-3}$, the curve of the DPL calculated for the value $\tau_1 = 30 \times 10^{-5}$. The figure shows a pronounced effect for the cooling parameter, where the maximum of the temperature does not appear at the irradiated surface like the previous figure, but it is shifted into the medium. According to [30], for the LS and CTE models in the case of volumetric absorption, the value of the temperature at the irradiated surface and at the location where the maximum occurs is smaller than the case of surface absorption, while the penetration and the maximum locations do not affect the mechanism of heating. This is because of the greater thermal expansion in the case of the volumetric absorption than the surface absorption.

Figure 7 represents the spatial distribution of the temperature of the DPL model calculated for $h = 0$ at $t = 4 \times 10^{-3}$ and different values of τ_1 at a fixed value of τ_2. It is observed that as τ_1 approaches the value of τ_2, the penetration into the medium increases and as it takes values farther from τ_2 the gradient of temperature decreases in a region closes to the surface and then increases after that. This figure, much like Figure 5, agrees with Figure 3 from where the behavior goes towards the generalized or classical coupled theories.

Figure 8 represents the spatial distribution of the temperature of the DPL model calculated for $h \neq 0$ at $t = 4 \times 10^{-3}$ and different values of τ_1 at a fixed value of τ_2. The effect of the cooling is evidently appearing where the maximum temperature is shifted towards greater z-values. Furthermore, this figure is like Figure 6.

Figure 9 illustrates the spatial distribution of the displacement w of the DPL model calculated for $h = 0$ and $\tau_1 = 30 \times 10^{-5}$ at several times. The curves show that w appears in a region close to the irradiated surface; this behavior is due to the delayed waves of the displacement, which come as a result of the heating process. Negative signs are characterizing the displacement, which is due to the geometry of the target, where the positive direction is pointed into the medium, while the displacement grows in the reverse direction. The figure shows a pronounced increase in both the magnitude and the penetration into the medium, with increasing time; this is due to the increased penetration of the temperature with time.

Figure 10 represents the spatial distribution of w for three different models, namely, LS, DPL and CTE; the curves calculated for $h = 0$ and $t = 4 \times 10^{-3}$, the curve of the DPL model calculated for the value $\tau_1 = 30 \times 10^{-5}$. The figure shows that the value at the irradiated surface is greater for LS and DPL models than the CTE model while the penetration does not affect; this behavior can be interpreted from Figure 5 where the gradient of the temperature of LS and DPL models is smaller than the gradient of the CTE. According to [30], the curves of the CTE and LS for the surface absorption are greater than the case of the volumetric absorption, while the penetration is approximately equal.

Figure 11 represents w for three different models, namely LS, DPL and CTE, the curves calculated for $h \neq 0$ and $t = 4 \times 10^{-3}$, the curve of the DPL model calculated for the value $\tau_1 = 30 \times 10^{-5}$. The figure shows that due to cooling, the displacement of the three theories are practically the same. Furthermore, the penetration into the medium is slightly greater in the case of $h \neq 0$ than that of $h = 0$ in Figure 10. In the case of cooling, the case of surface absorption shows smaller magnitude and penetration than the case of volumetric absorption [30].

Figure 12 represents the spatial distribution of w of the DPL model calculated for different values of τ_1 at a fixed value of τ_2 at the time $t = 4 \times 10^{-3}$ with $h = 0$. It is noted from the figure that the displacement is increased with decreasing τ_1, while the penetration is not affected.

Figure 13 represents w of the DPL model calculated for different values of τ_1 at a fixed value of τ_2 at the time $t = 4 \times 10^{-3}$ with $h \neq 0$, from the figure, a very weak effect for τ_1 with the existence of cooling being observed, where the curves seem to coincide.

Figure 14 represents the spatial distribution of σ_{zz} of the DPL model calculated for $h = 0$ and $\tau_1 = 30 \times 10^{-5}$ at several times, beside a small figure representing σ_{zz} near the irradiated surface. According to Equation (13), the behavior of σ_{zz} is due to the effect of the gradient of the displacement and the temperature. The gradient of the displacement is practically has its effect in a region very close to the irradiated surface and its effect increases with time (see Figure 9). As z increases, the effect of the gradient of w vanishes and then the effect of the temperature appears evidently as in Figure 4. For the times ($t = 3 \times 10^{-3}$, $t = 3.4 \times 10^{-3}$) or before the temperature reaches its maximum, the stress does not possess a positive peak in the vicinity of the irradiated surface, the positive peak appears for ($t = 4 \times 10^{-3}$, $t = 5 \times 10^{-3}$).

Figure 15 represents σ_{zz} for three different models, namely LS, DPL and CTE, calculated for $h = 0$ and time $t = 4 \times 10^{-3}$, the curve of the DPL model calculated for the value $\tau_1 = 30 \times 10^{-5}$. In the vicinity of the irradiated surface, the positive peak of σ_{zz} is greater for the generalized theories (LS, DPL) than the CTE model. By increasing the z-value, the temperature effect becomes pronounced and shows the same behavior as in Figure 5. According to [30], both positive and negative peaks of the stress σ_{zz} are greater in the case of surface absorption than the case of the volumetric absorption for LS and CTE models.

Figure 16 represents σ_{zz} for three different models, namely LS, DPL and CTE, calculated for $h \neq 0$ and time $t = 4 \times 10^{-3}$, the curve of the DPL model calculated for the value $\tau_1 = 30 \times 10^{-5}$. This figure agrees with Figures 6 and 11, where in a region closest to the irradiated surface, the cooling has a slight effect on displacement and so its gradient. By increasing the z-value, the behavior of the temperature appears with the cooling effect. The figure agrees with the previous one in terms of the effect of the technique of absorption [30].

Figure 17 represents the spatial distribution of σ_{zz} of the DPL model calculated for different values of τ_1 at a fixed value of τ_2 at the time $t = 4 \times 10^{-3}$ with $h = 0$. Close to the irradiated surface, the positive peak increases with a decreasing value of τ_1. By increasing the z-value σ_{zz} shows the same behavior of the temperature for the same case (Figure 7).

Figure 18 represents σ_{zz} of the DPL model calculated for different values of τ_1 at a fixed value of τ_2 at time $t = 4 \times 10^{-3}$ with $h \neq 0$. In the vicinity of the irradiated surface, the stresses seem to coincide with each other, which is due to the cooling effect. By increasing the z-value, the behavior of the temperature with cooling appears evidently.

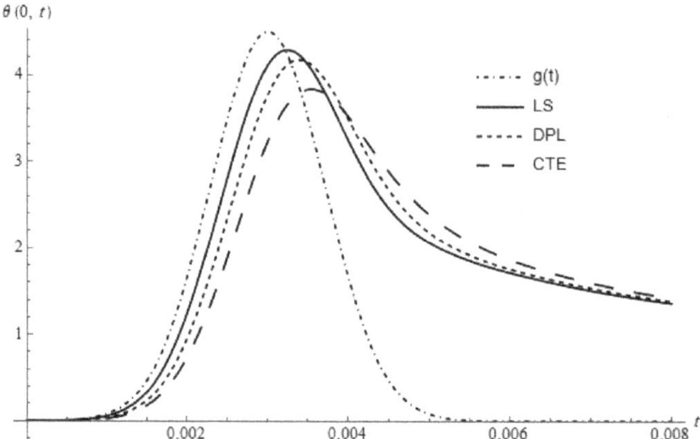

Figure 1. The irradiating laser pulse profile $g(t)$ and the time dependence of the surface temperature $\theta(0, t)$ calculated for three different theories for $h = 0$.

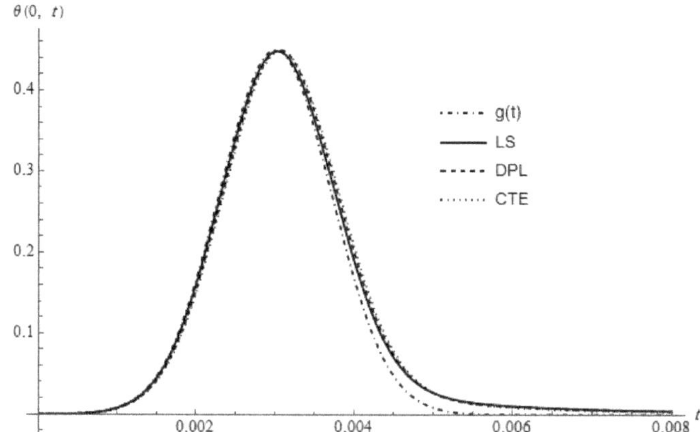

Figure 2. The irradiating laser pulse profile $g(t)$ and the time dependence of the surface temperature $\theta(0, t)$ calculated for three different theories for $h \neq 0$.

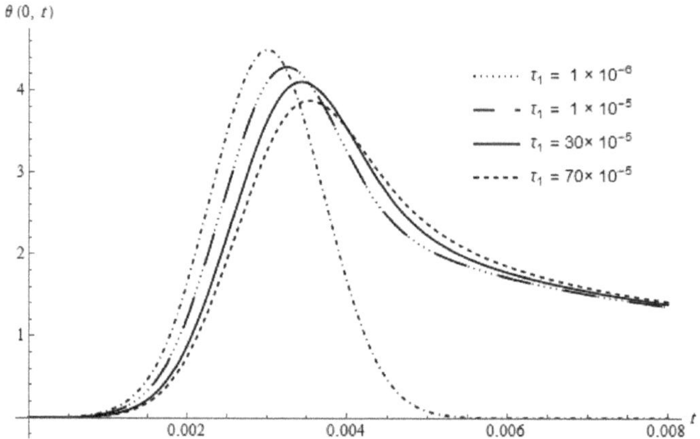

Figure 3. The irradiating laser pulse profile $g(t)$ and the time dependence of the surface temperature $\theta(0, t)$ of the DPL theory calculated for $h = 0$ and different values of τ_1.

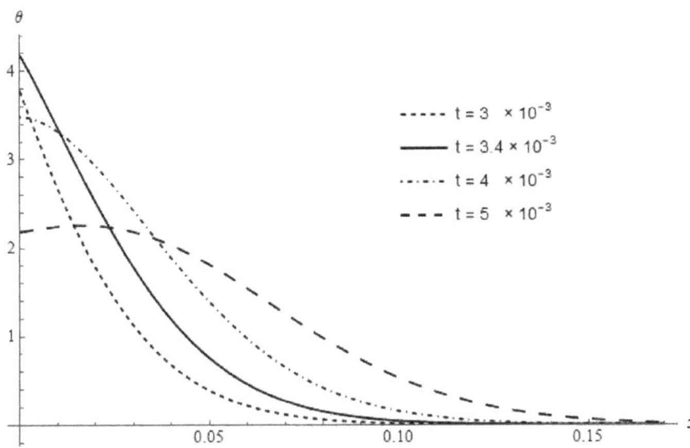

Figure 4. The temperature $\theta(z, t)$ of the DPL theory as a function of z calculated for $h = 0$ at different times.

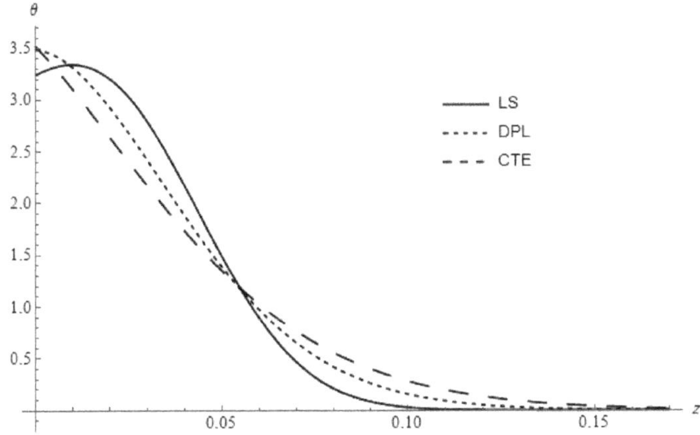

Figure 5. The temperature $\theta(z, t)$ of three theories as a function of z calculated for $h = 0$ at $t = 4 \times 10^{-3}$.

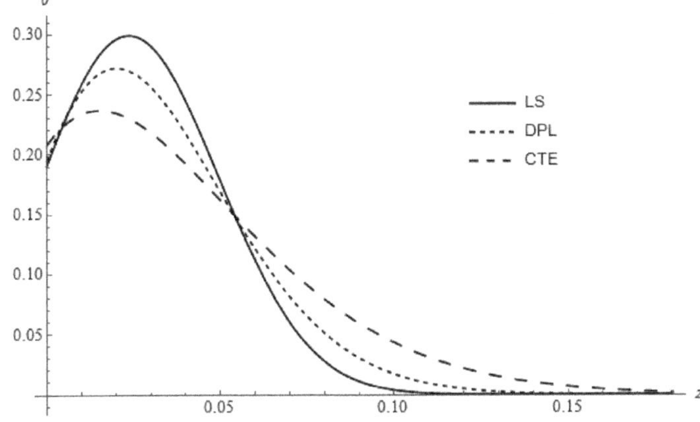

Figure 6. The temperature $\theta(z, t)$ of three theories as a function of z calculated for $h \neq 0$ at $t = 4 \times 10^{-3}$.

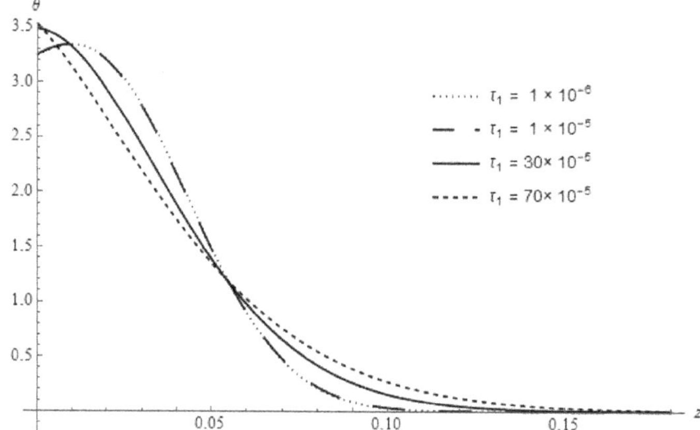

Figure 7. The temperature $\theta(z, t)$ of the DPL theory as a function of z calculated for $h = 0$ and different values of τ_1 at $t = 4 \times 10^{-3}$.

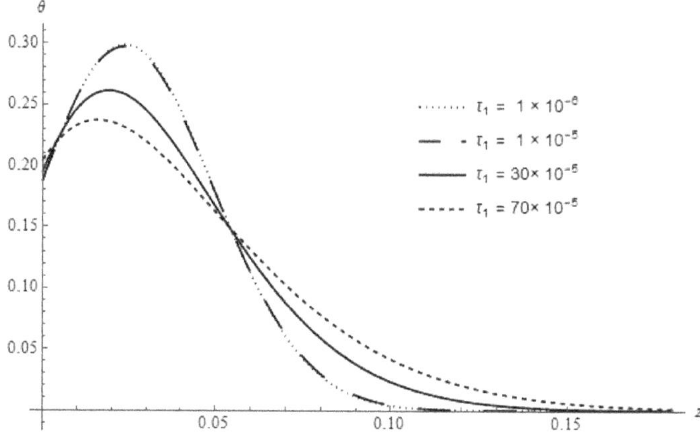

Figure 8. The temperature $\theta(z, t)$ of the DPL theory as a function of z calculated for $h \neq 0$ and different values of τ_1 at $t = 4 \times 10^{-3}$.

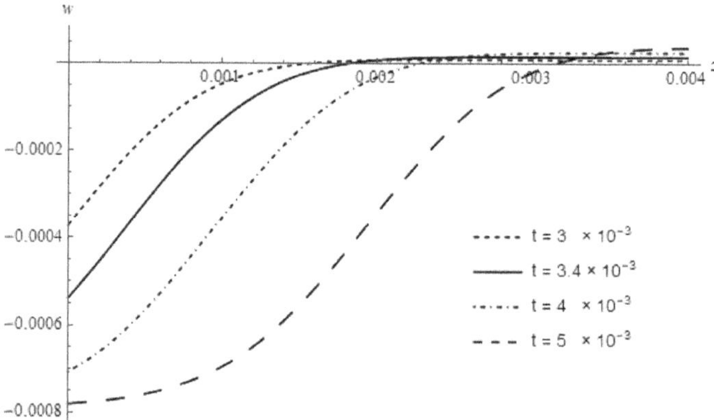

Figure 9. The displacement $w(z, t)$ of the DPL theory as a function of z calculated for $h = 0$ at different times.

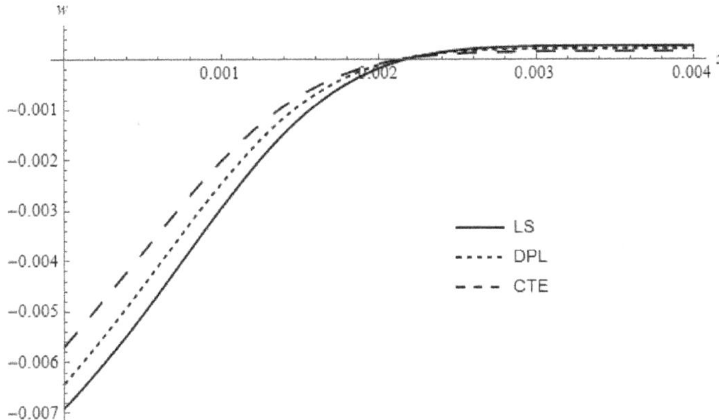

Figure 10. The displacement $w(z,t)$ of three theories as a function of z calculated for $h = 0$ at $t = 4 \times 10^{-3}$.

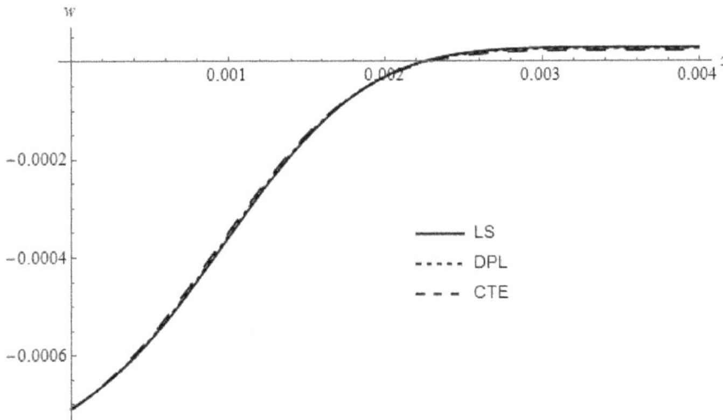

Figure 11. The displacement $w(z,t)$ of three theories as a function of z calculated for $h \neq 0$ at $t = 4 \times 10^{-3}$.

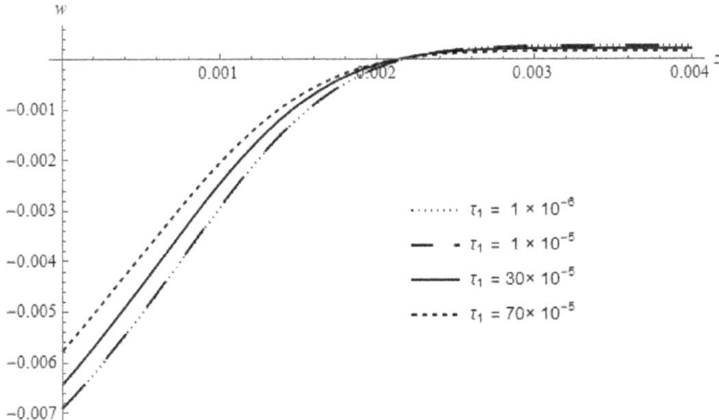

Figure 12. The displacement $w(z,t)$ of the DPL theory as a function of z calculated $h = 0$ and different values of τ_1 at $t = 4 \times 10^{-3}$.

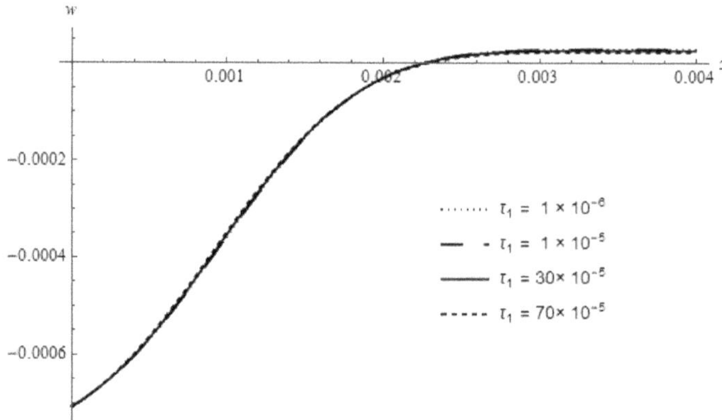

Figure 13. The displacement $w(z,t)$ of the DPL theory as a function of z calculated $h \neq 0$ and different values of τ_1 at $t = 4 \times 10^{-3}$.

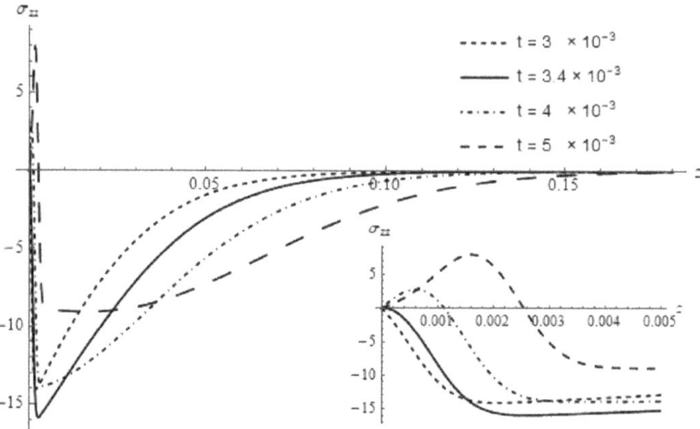

Figure 14. The stress $\sigma_{zz}(z,t)$ of the DPL theory as a function of z calculated for $h = 0$ at different times.

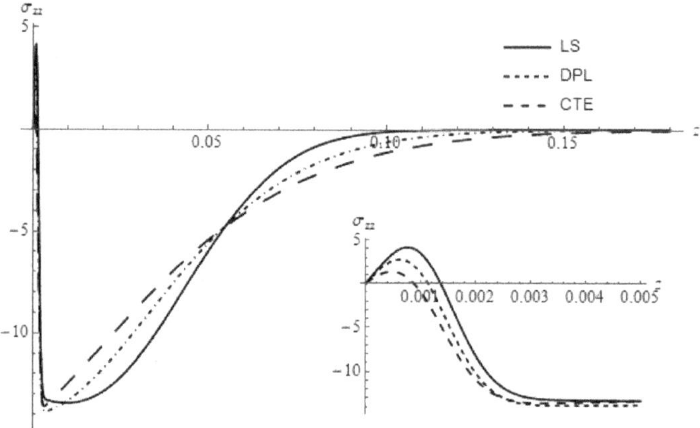

Figure 15. The stress $\sigma_{zz}(z,t)$ of three theories as a function of z calculated for $h = 0$ at $t = 4 \times 10^{-3}$.

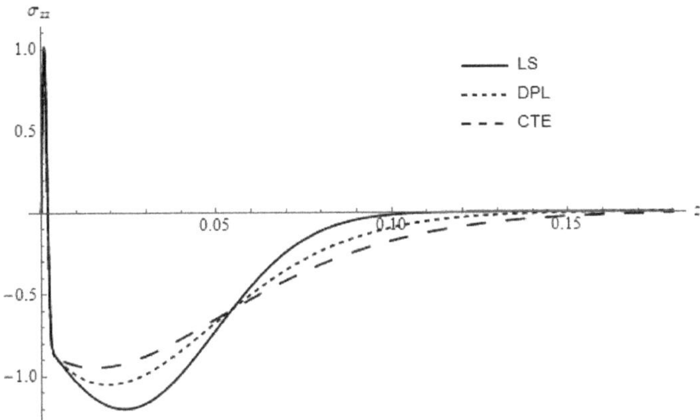

Figure 16. The stress $\sigma_{zz}(z,t)$ of three theories as a function of z calculated for $h \neq 0$ at $t = 4 \times 10^{-3}$.

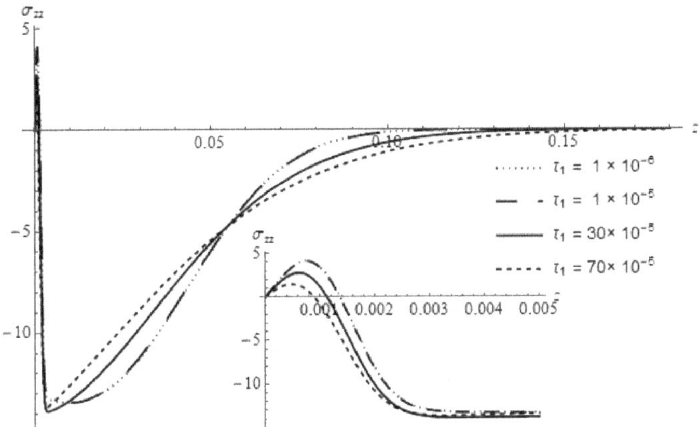

Figure 17. The stress $\sigma_{zz}(z,t)$ of the DPL theory as a function of z calculated $h = 0$ and different values of τ_1 at $t = 4 \times 10^{-3}$.

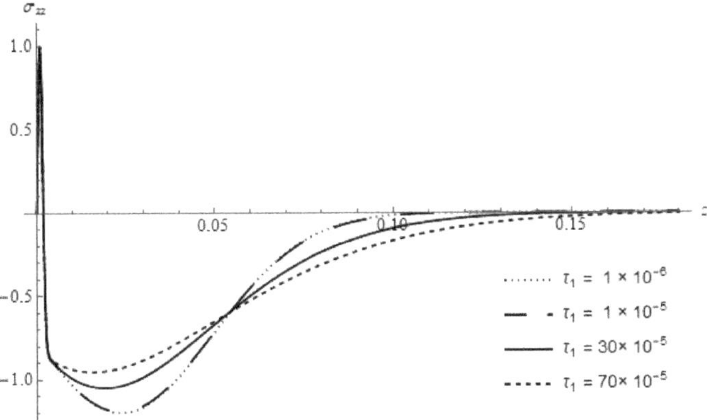

Figure 18. The stress $\sigma_{zz}(z,t)$ of the DPL theory as a function of z calculated $h \neq 0$ and different values of τ_1 at $t = 4 \times 10^{-3}$.

7. Conclusions

This paper was devoted to study the thermoelastic response induced by volumetric absorption of a uniform laser radiation in a homogeneous and isotropic thermoelastic half-space whose surface is exposed to heat losses. From the above, it can be concluded that:

1. The responses induced by the surface absorption are greater than the responses induced by the volumetric absorption, which loses some energy in absorbing the radiation throughout the medium.
2. The results obtained by employing the generalized theories are not in contradiction with the well-known physical phenomena, while the results obtained from the CTE model display its nature evidently.
3. In both surface and volumetric absorption, the cooling parameter shows a very slight effect on the displacement and its gradient, while it shows a pronounced effect on the temperature, and it can control the influence of the laser power.
4. The response of the DPL model appears as if it were a case between the LS and the CTE models.
5. A clear effect is seen on all studied fields for the phase lag of the temperature gradient at a specified value of the phase lag of the heat flux.

Acknowledgments: The author would like to thank Deanship of Scientifc Research at Majmaah University for supporting this work under project number [R-1441-82].

Conflicts of Interest: The author declares no conflict of interest.

References

1. Biot, M.A. Thermoelasticity and irreversible thermodynamics. *J. Appl. Phys.* **1956**, *27*, 240–253. [CrossRef]
2. Hetnarski, R.B. Coupled one-dimensional thermal shock problem for small times. *Arch. Mech. Stosow.* **1961**, *13*, 295–306.
3. Hetnarski, R.B. Solution of the coupled problem of thermoelasticityi in the form of series of functions. *Arch. Mech. Stosow.* **1964**, *16*, 919–941.
4. Boley, B.A.; Tolins, I.S. Transient coupled thermoelastic boundary value problems in half-space. *J. Appl. Mech.* **1962**, *29*, 637–646. [CrossRef]
5. Lord, H.W.; Shulman, Y. A generalized dynamical theory of thermoelasticity. *J. Mech. Phys. Solids* **1967**, *15*, 299–309. [CrossRef]
6. Ignaczak, J. A uniqueness theorem for stress-temperature equations of dynamic thermoelasticity. *J. Therm. Stresses* **1978**, *1*, 163–170. [CrossRef]
7. Ignaczak, J. A note on uniqueness in thermoelasticity with one relaxation time. *J. Therm. Stresses* **1982**, *5*, 257–263. [CrossRef]
8. Sherief, H.; Dhaliwal, R. Generalized one-dimensional thermal shock problem for small times. *J. Therm. Stresses* **1981**, *4*, 407–420. [CrossRef]
9. Sherif, H. Fundamental solution of the generalized themoelastic problem for short times. *J. Therm. Stresses* **1986**, *9*, 151–164. [CrossRef]
10. Sherief, H.; Hamza, F. Generalized thermoelastic problem of a thick plate under axisymmetric temperature distribution. *J. Therm. Stresses* **1994**, *17*, 435–452. [CrossRef]
11. Green, A.E.; Lindsay, K.A. Thermoelasticity. *J. Elast.* **1972**, *2*, 1–7. [CrossRef]
12. Green, A.E.; Naghdi, P.M. Thermoelasticity without energy dissipation. *J. Elast.* **1993**, *31*, 189–208. [CrossRef]
13. Hetnarski, R.B.; Ignaczak, J. Soliton-like waves in a low temperature nonlinear thermoelastic solid. *Int. J. Eng. Sci.* **1996**, *34*, 1767–1787. [CrossRef]
14. Ozisik, M.N.; Tzou, D.Y. On the wave theory of heat conduction. *J. Heat Transf. (ASME)* **1994**, *116*, 526–535. [CrossRef]
15. Tzou, D.Y. A unified field approach for heat conduction from macro-to micro-scales. *J. Heat Transf.* **1995**, *117*, 8–16. [CrossRef]
16. Tzou, D.Y. Experimental support for the lagging behavior in heat propagation. *J. Thermophys. Heat Transfer* **1995**, *9*, 686–693. [CrossRef]
17. RoyChoudhuri, S.KR. One-dimensional thermoelastic waves in elastic half-space with dual phase-lag effects. *J. Mech. Mater. Struct.* **2007**, *2*, 489–503. [CrossRef]

18. Abouelregal, A. A problem of a semi-infinite medium subjected to exponential heating using a dual-phase-lag thermoelastic model. *Appl. Math.* **2011**, *2*, 619–624. [CrossRef]

19. El-Karamany, A.S.; Ezzat, M.A. On the Dual-Phase-Lag thermoelasticity theory. *Meccanica* **2014**, *49*, 79–89. [CrossRef]

20. Abbas, I.; Zenkour, A.M. Dual-Phase-Lag model on thermoelastic interactions in a semi-infinite medium subjected to ramp-type heating. *J. Comput. Theor. Nanos.* **2014**, *11*, 642–645. [CrossRef]

21. Marin, M.; Broadbridge, P.; Öchsner, A. Well-posed dual-phase-lag model of a thermoelastic dipolar body. *ZAMM Z. Angew. Math. Mech.* **2017**, *97*, 1645–1658. [CrossRef]

22. Hassan, A.F.; El-Nicklawy, M.M.; EL-Adwi, M.K.; Nasr, E.M.; Hemida, A.A.; El-Ghaffar, O.A.A. Heating effects induced by a pulsed laser in a semi-ifinite target in view of the theory of linear systems. *Opt. Laser Technol.* **1996**, *28*, 337–343. [CrossRef]

23. McDonald, F.A. On the precursor in laser-generated ultrasound waveforms in metals. *Appl. Phys. Lett.* **1990**, *56*, 230–232. [CrossRef]

24. Dubois, M.; Enguehard, F.; Bertrand, L. Modeling of laser thermoelastic generation of ultrasound in an orthotrobic medium. *Appl. Phys. Lett.* **1994**, *64*, 554–556. [CrossRef]

25. Wang, X.; Xu, X. Thermoelastic wave induced by pulsed laser heating. *Appl. Phys. A* **2001**, *73*, 107–114. [CrossRef]

26. Henain, E.F.; Hassan, A.F.; Megahed, F.; Tayel, I.M. Thermo-elastic half space under illumination of a laser beam by using Lord and Shulman theory. *J. Therm. Stresses* **2014**, *37*, 51–72. [CrossRef]

27. Yossef, H.M.; El-Bary, A.A. Thermoelastic material response due to laser pulse heating in context of four theorems of thermoelasticity. *J. Therm. Stresses* **2014**, *37*, 1379–1389. [CrossRef]

28. Allam, M.N.M.; Tayel, I.M. Generalized thermoelastic functionally graded half space under surface absorption of laser radiation. *J. Theor. Appl. Mech.* **2017**, *55*, 155–165. [CrossRef]

29. Abbas, I.A.; Marin, M. Analytical solution of thermoelastic interaction in a half-space by pulsed laser heating. *Phys. E Low-Dimens. Syst. Nanostruct.* **2017**, *87*, 254–260. [CrossRef]

30. Tayel, I.M.; Hassan, A.F. Heating a thermoelastic half space with surface absorption pulsed laser using fractional order theory of thermoelasticity. *J. Theor. Appl. Mech.* **2019**, *57*, 489–500. [CrossRef]

31. Debnath, L.; Bhatta, D. *Integral Transforms and Their Applications*; Taylor and Francis: New York, NY, USA, 2015.

© 2020 by the author. Licensee MDPI, Basel, Switzerland. This article is an open access article distributed under the terms and conditions of the Creative Commons Attribution (CC BY) license (http://creativecommons.org/licenses/by/4.0/).

Article

Experiment Study of Rapid Laser Polishing of Freeform Steel Surface by Dual-Beam

Yongquan Zhou *, Zhenyu Zhao, Wei Zhang, Haibing Xiao and Xiaomei Xu

School of Intelligent Manufacturing and Equipment, Shenzhen Institute of Information Technology, Shenzhen 518172, China; zhaozy@sziit.edu.cn (Z.Z.); zhangwei@sziit.edu.cn (W.Z.); xiaohb@sziit.edu.cn (H.X.); xuxm@sziit.edu.cn (X.X.)
* Correspondence: zhouyq@sziit.edu.cn; Tel.: +86-755-89226591

Received: 28 April 2019; Accepted: 13 May 2019; Published: 16 May 2019

Abstract: One of the challenges regarding widespread use of parts made from alloy steel is their time-consuming polishing process. A rough freeform surface of part has been often expected to be polished rapidly up to a smooth surface finish. The focus of this study is to develop a fast polishing method of freeform surface by using dual-beam lasers. The dual-beam laser system consists of continuous laser (CW) and pulsed laser based on a five-axis CNC device. In this study, a series of experiments of CW laser polishing present the effects of different spot irradiation on surface topography, then the combination trajectory of zigzag and square waveform of pulsed laser is explored to realize a "melting peak for filling into valley" (MPFV) method. The polishing experiment on a semisphere of S136H steel polished by dual-beam shows that a rough semisphere surface was rapidly polished from initial state value of Sa (=877 nm) to post-polished value of Sa (=142 nm), and the polishing efficiency is as high as 2890 cm^2/H.

Keywords: dual-beam; beam shaper; MPFV method; laser polishing; zigzag-square wave

---

1. Introduction

Freeform surface plays significant role of enhancing functional and aesthetical properties of products. The increasing demand for parts with free-form surfaces in aerospace, communications, energy, mold making, kitchenware and medicine requires that polishing technology provides excellent surface finish in high efficiency. Current automated polishing techniques often cannot be used on parts with freeform surfaces and functional relevant edges, therefore the finishing for these parts is often done manually [1,2]. Due to the low polishing efficiency (typically maximum 30 cm^2/H from initial state value of Sa (=877 nm) to post-polished value of Sa (=142 nm) by a skilled worker, it is often expected that the time consuming polishing process should be replaced by an innovative rapid one.

One new approach to automate this work is rapidly polishing by means of laser radiation. This is why, in recent years, laser polishing has become more widely used due to the ability to quickly and efficiently polish surfaces [3]. Besides metal, laser is able to polish non-metal materials, ultrafast lasers have been used to produce complex shapes in glass [4] while CW lasers have been used for polishing them with extremely low surface roughness [5]. In polishing process, laser is melting a thin surface layer, melted material flows from the peaks to the valleys due to the surface tension, to evenly distribute the recently melted material across the surface and thus creating a much smoother surface finish [6]. Typical roughness after laser polishing are in the range of Sa (150–512 nm), which meets the medium demand on surface quality of a large number of parts in many industrial fields.

Most researchers are focusing on the study of laser polishing of flat surface of workpiece; however, due to the existence of difference of the surface tension of melted pool between flat surface and freeform surface, a new approach to attenuate longer spatial wavelengths of freeform surface is often asked for.

2. CW Laser Polishing

2.1. CW Laser Beam Profile Pattern

The dual-laser polishing system consists of CW laser and pulsed laser. First of all, the performance of CW laser polishing should be studied. Laser polishing differs from more widely known techniques of laser engraving and laser ablation in that it is an equating process, no adding material or removing material as in those techniques. The profile pattern of CW laser beam from the laser generator is Gaussian beam instead of top-hat beam, where the spot is circular but with a uniform intensity profile [7]. A Gaussian beam results in a non-uniform process that polishes some areas of the panel more effectively than others, due to the differences in their thermal cycles. To avoid any ablation on polished surface, the laser beam must be a top-hat beam rather than a Gaussian beam, so that the power density of the spot can be uniform to ensure no any material is vaporized in the molten pool. Meanwhile, there are several parameter factors to influence the surface quality during CW laser polishing. Previous works respectively explored using the parameters such as power density, scanning speed, scanning line step-over, scan trajectories during CW laser polishing for improved surface finish [1,6,8,9].

The spot irradiation of CW laser is a critical factor in CW laser polishing process; however, its effect was rarely introduced in previous works since it was often fixed. In the experiment study, as shown in Figure 1, a combined system consisting of a reverse beam expander and a beam shaper was established, which does not just convert a Gaussian beam into a top-hat beam, but also realizes stepless adjustment of spot diameter from 0.32 to 0.54 mm, thus realizing the adjustment of the spot irradiation. Ideally, the smaller the M^2 value of initial CW Gaussian beam, the more uniform the distribution of spot irradiation of converted top-hat beam.

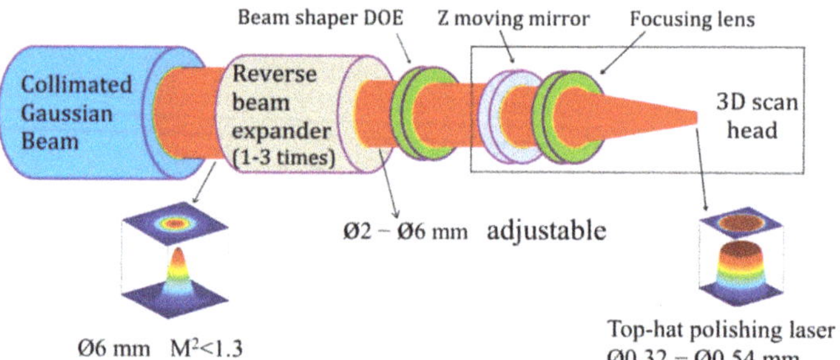

Figure 1. Combination of CW laser expander and shaper.

High quality of the beam profile of CW laser plays a key role in the process optimization. To portray the degree of uniform intensity of beam profiles, three patterns of top-hat beam profile are marked by "Ideal", "Medium" and "Poor" respectively, and given in Figure 2 regarding to the setup of diameter ø0.47 mm of the top-hat beam. The "Ideal" beam pattern has a lowest irregularity figure, in order not to introduce wave-front errors which would degrade the beam shaper performance. The quality of top-hat beam profile is depended on many factors such as mirror's flatness specification, M^2 value of input Gaussian beam, the laser waist position where the beam shaper element works [10]. The degree of irregularity increases from "Medium" beam pattern to "Poor" one since some regions overheat compared to other regions.

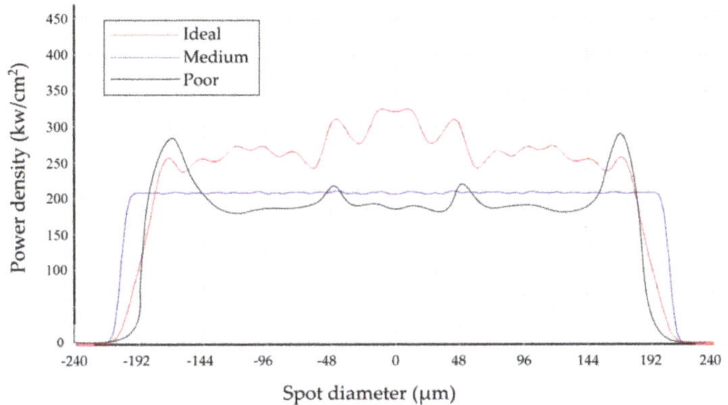

Figure 2. Quality of top-hat beam profile.

2.2. CW Laser Polishing Using Different Sopt Diameters

A large number of cores/cavities of translucent plastic injection mold are made of S136H steel, the surface of cores/cavity often needs to be polished to Sa < 190nm or even lower. Thus, the first focus of this study will be to generate more comprehensive understanding of surface quality of CW laser polishing S136H steel with different spot irradiation. Four groups of specimens of ground flat S136H tool steel (10 mm thickness) were prepared. Each group had 9 specimens with different initial roughness, ranging from Sa (=768 nm) to Sa (=4826 nm). To portray the effects that the beam spot irradiation have on area polishing, a rough polishing experiment was conducted using a CW laser by the adjustment of spot irradiation from 262–746 kw/cm^2 based on the reverse beam expander as shown in Figure 1. Each group was polished with different spot irradiation as shown in Table 1.

Table 1. Group number corresponding to different spot irradiation.

Group #	I	II	III	IV
Spot diameter Ds (mm)	0.54	0.47	0.40	0.32
Spot Irradiation Ir. (kw/cm^2)	262	346	477	746

Except the spot irradiation and scanning speed, the surface polishing strategy was similar to previous works [11–13], and the polishing parameters were optimized and given in Table 2. To polish rapidly, the scanning speed was setup over 600% higher than previous works [3,14].

Table 2. Optimized parameters of CW laser polishing in each group.

Factor Name	Optimized Value or Feature								
Roughness of initial state Sa (nm)	768	983	1293	1862	2443	3020	3612	4361	4826
Scanning speed (mm/s)	800	800	750	750	700	700	650	650	600
Top-hat beam profile	Between Medium and Ideal								
Polishing parameters	Wavelength: 1080 nm, Power: 600 W, Step-over: 0.1 mm								

Total 36 specimens were polished, each specimen was polished more than 3 times in different regions in order to make an error bar of roughness measurement. The surface measurements were taken of the ground and laser polished surfaces using a white light interferometer (BRUKER WYKO Contour GT-K, Billerica, MA, USA). Figure 3 shows the roughness of the laser polished surfaces corresponding to each initial rough ground flat surface with the varying spot irradiation. Point C's error bar with scale is shown in Figure 2 to reveal the difference of roughness measured by polishing in different regions of a specimen. Point A and Point B present the minimum roughness and maximum roughness obtained among the 36 polished specimens and the results are shown in Figures 3 and 4.

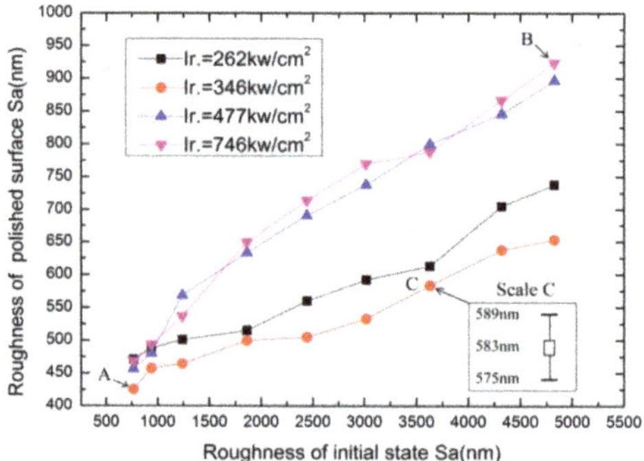

Figure 3. Polished roughness related to initial roughness.

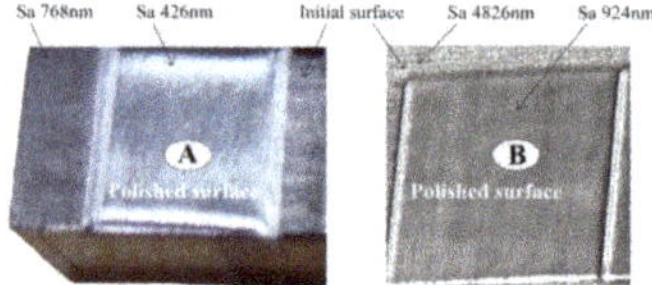

Figure 4. Polished specimens.

The CW laser polishing process was able to achieve a surface roughness reduction from 44%–81%, resulting in the areal surface roughness dropping from its initial state value of *Sa* (=4826 nm) to a final post-polished value of *Sa* (=924 nm).

Figure 3 reveals the fact that an appropriate spot irradiation is one of key factors to influence the surface finish when CW laser is polishing a steel surface in high efficiency. Obviously, 346 kw/cm² is the optimized value of spot irradiation in the experiment. The ablation or evaporation occurred in the melted pool when the spot irradiation is over 477 kw/cm², so that the roughness of polished surface is higher than that of polished surface on which the spot irradiation (≤346 kw/cm²) was applied.

3. Experiment of Dual-Beam Laser Polishing

In laser polishing process, many aspects of the research surrounding the topic remain focused on process optimization to achieve finish surface. Often, when experimenting with the optimization of laser polishing parameters (power density, scanning speed, spot diameter, etc.), a so-called "melting peak for filling into valley" (MPFV) method is implemented. The MPFV method consists of melting a series of peaks of topography of surface independent of one another, each flowing into adjacent valleys by surface tension of melted pool.

Nüsser et al. [15] has developed a dual-laser system in which the CW laser was to pre-heat the surface as high as possible without melting the surface, the pulsed laser implemented MPFV method to polish the surface.

In the experiment study, the CW laser not only pre-heats the surface but also roughly polishes the surface. To achieve this, experiments will be conducted using MPFV method with dual-beam lasers and an innovation of trajectory of pulsed laser will be analyzed.

Hafiz et al. [8] and Nüsser et al. [15] have investigated that a combined CW and pulse laser polishing as a two-step process on various alloy melts and shown to improve the surface finish. However, such previous works focused on the polishing of ground flat surface rather than freeform surface. In the investigation, a dual-beam laser polishing system has been developed to implement the MPFV method for freeform surface.

3.1. Dual-Beam Laser Polishing Device

A diagram of the experimental setup of the dual-beam laser polishing system is given in Figure 5. The system is composed of two 3D scan heads and a 2-axis CNC rotation table. A CW laser beam, which is reshaped from Gaussian beam to top-hat beam (Figure 1), enters a Z moving lens and focusing lens of the first 3D scan head. A pulsed laser beam with top-hat profile pattern from a laser generator directly enters a Z moving lens and focusing lens of the second 3D scan head. Taking for example of first 3D scan head, after moving lenses, the CW laser beam diverges rapidly until it enters one or two focusing lenses. The beam, now converging, passes through and is directed by a set of X and Y mirrors moved. The orthogonal arrangement of the X and Y mirrors direct the beam down towards and over the length and width of the working field. The focusing height of laser is adjusted by moving Z lens according to the Z coordinates of 3D surface model. The 3D freeform surface is activated by the laser beam, the maximum angle of incidence of the laser beam on the surface to be machined must not be exceeded if safe activation is to be achieved. The angle of incidence is the angle between the orthogonal to the activated surface and the laser beam. Usually the maximum angle of incidence is no bigger than 30°, so that some tilted surfaces are out of its machining scope, therefore a 2D rotation table is needed to make sure any tilted surface could be machined by its rotation. The system is able to move internal laser beam in three primary axes, designated $\vec{X}, \vec{Y}, \vec{Z}$, and another two axes created by the rotation of the x and y-axes, designed $\overset{\frown}{A}, \overset{\frown}{C}$, therefore five CNC axes are $\vec{X}, \vec{Y}, \vec{Z}, \overset{\frown}{A}, \overset{\frown}{C}$ respectively. Figure 6 is the five-axis CNC device with dual-beam system. The motion in the five axes should be synchronized in order to achieve predictable 3D polishing trajectories of the beam focus.

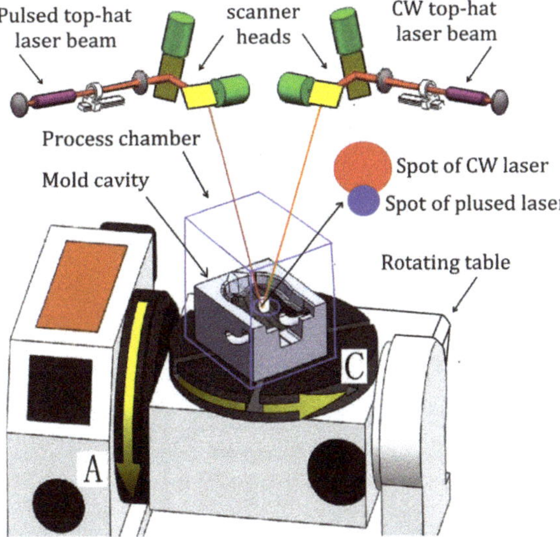

Figure 5. Dual-beam laser polishing system.

The above experiment indicates that CW laser polishing is of high polishing efficiency since the scanning speed is over 800 mm/s, but the polished surface roughness did not reach to $Sa < 400$ nm. To achieve the advantage of both high polishing efficiency and smoother surface finish, pulsed laser micro polishing (PLμP) is conducted to implement the MPFV Method.

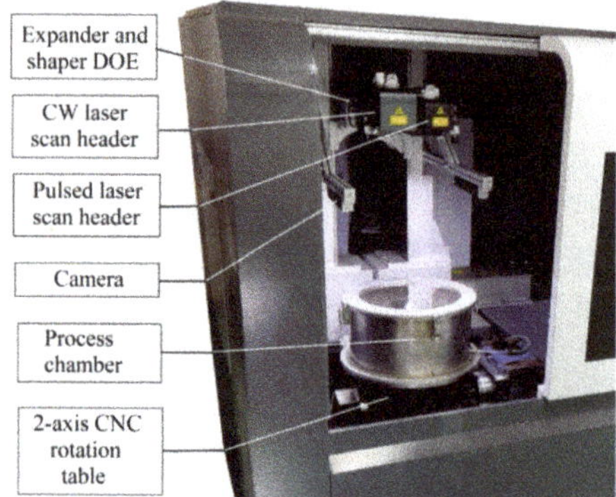

Figure 6. Five-axis CNC device with dual-beam.

Figure 7 is the topography of the specimen marked "A" in Figure 4. Ukar et al. [16] showed that due to a series of surface irregularities in the melted pool, there are zones with greater amounts of material, which when melted generate areas of higher surface tension. This surface tension tends to eliminate the irregularities within the melted area, producing material displacement into the melt pool. In other words, the surface tension becomes depreciated, resulting in the poor performance of eliminating the irregularities during CW laser polishing process with high scanning speed (800 mm/s) in that peak region shown in Figure 7 is not capable of fully filling into the valleys adjacent and surrounding to itself, therefore the melted peak is cooled and solidified rapidly during MPFV process, forming a new series of smaller peaks (red areas in Figure 7).

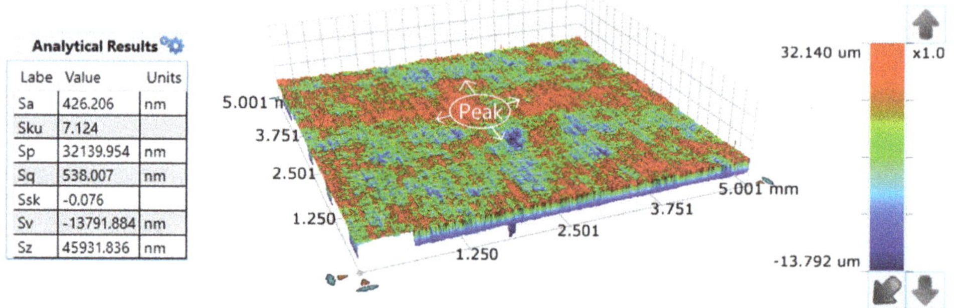

Figure 7. Topography of polished surface by CW laser and measured data.

3.2. Trajectory of Zigzag-Square Wave of Dual-Beam

PLμP is capable of achieving significant surface smoothing quickly without removing material, and promotes the migration of molten pool materials, since it melts a metallic surface and allows surface tension effects to smooth the surface [17]. PLμP is a helpful tool to implement the MPFV method. In dual-beam laser polishing system, the pulsed laser follows up with the CW laser closely with a certain level of overlap. The new peak which is formed during CW laser polishing process, will not be solidified immediately since the pulsed laser beam is following up with it and making the new peak re-melted instead. A very smaller layer of material creating a smaller molten pool, which uses the physics of surface tension, evenly distributes the recently re-melted material across the surface, and creates much smoother surface finish.

Figure 7 indicates that the peaks formed during CW laser polishing exist in both transverse and longitudinal directions. To improve the MPFV performance of PLμP, the pulsed laser should alternately move along the transverse and longitudinal directions in order to mitigate the peaks to adjacent valleys in the two directions. Therefore a combination trajectory of zigzag and square waveform of the pulsed laser is proposed, zigzag is the path for pulsed laser to follow up with CW laser, and the square wave is the path for pulsed laser to mitigate the peaks to adjacent valleys in the transverse and longitudinal directions. The trajectories of dual-beam are figured out in Figure 8. Δ_1 is the step-over of the CW laser, Δ_2 is that of the pulsed laser, and L_O is the overlap between both beams.

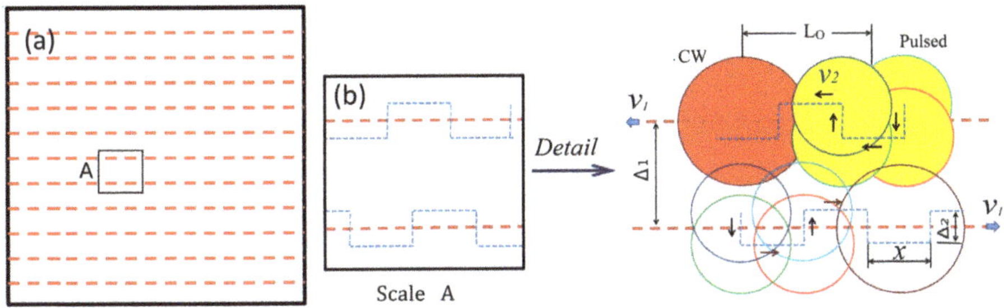

Figure 8. Trajectory of two lasers: (**a**) zigzag of CW laser (**b**) square-wave of pulsed laser.

The trajectory of CW laser is zigzag [3], and that of pulsed laser is the combination of zigzag and square wave. x is a half wavelength of the square wave. Nüsser et al. [15] has concluded that an appropriate overlap L_O led to the smallest micro roughness, so the range of $2x \leq L_O \leq 3x$ was set up in the experiment, excessive L_O value makes the new peak solidified in advance of the pulsed laser working and weakens the surface tension of melted pool.

Both scanning speeds must match with each other to keep the overlap L_O in above range. If the scanning speed of CW laser is v_1 that of pulsed laser is v_2, then,

$$(1 + \frac{2\Delta_2}{3x})v_1 \leq v_2 \leq (1 + \frac{\Delta_2}{x})v_1 \tag{1}$$

The zigzag trajectory is a common practice in laser processing [3,18]; however, the pulsed laser is following up with the CW laser with its own square wave route besides the zigzag trajectory, its route network should be optimized individually.

According to the route network optimization of previous works [19,20], the offset contours of laser scanning trajectory are marked in four directions, which are from left to right, from right to left, from top to bottom and from bottom to top, and the endpoints and intersections are connected in turn. Both the endpoints and intersections of the contour form the vertices V_i of the route network, both the length of the line segment and the arc between the connections form the arc length A_i. A laser polishing track is connected from the beginning to the end. Based on the Dijkstra algorithm [20,21], V_{11}, V_{12}, V_{13}, V_{23}, V_{22}, V_{32} are the effective path vertices, which form the zigzag trajectory shown in Figure 9.

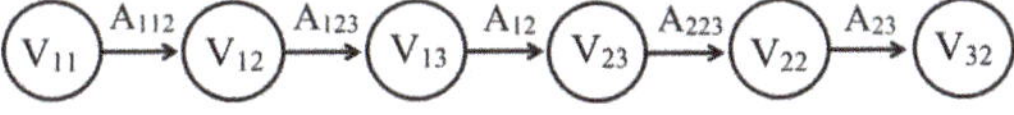

Figure 9. Zigzag trajectory.

According to the isometric migration method of zigzag strategy [19,22], the isometric migration value is the step-over Δ_1. All the isometric migration trajectories are connected in turn according to the shortest path of the network, and the direction-parallel path method of zigzag trajectory [22,23], the total length of the whole route can be obtained by accumulating the length of whole line segment, arc length A_{112}, A_{123}, A_{12}, ⋯, etc.

The trajectory of pulsed laser is square-wave, keeping a certain level of overlap L_O with the CW laser, V_{P11}, V_{P12}, V_{P13}, V_{P14}, ⋯, V_{P21}, V_{P22}, V_{P23}, V_{P24}, ⋯are the corresponding vertices of the square wave in route network optimization (Figure 10a). The distance between adjacent vertices is equal, the pulsed laser needs to traverse all vertices. According to the shortest path of the network and Dijkstra algorithm, the path graphs of trajectory of square wave of the pulsed laser (Figure 10b) is established by connecting these vertices in turn.

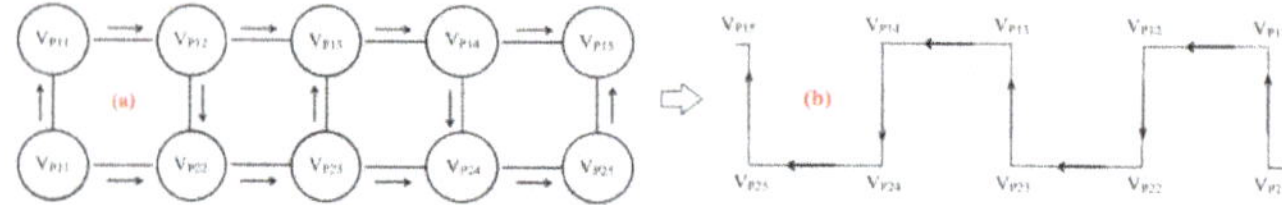

Figure 10. Trajectory optimization. (**a**) vertices of the square wave; (**b**) trajectory of square wave.

3.3. Experiment of Dual-Laser Beam Polishing

A semisphere of S136H steel which was machined by a CNC miller with the final post-machined roughness value of *Sa* (=877 nm) was prepared for the experiment of dual-beam polishing. The model of CW laser generator and pulsed laser generators were MFSC-700W plus and MFPT-120W plus MOPA (Manufacturer: Max, Shenzhen, China) respectively. The value of half wavelength *x* determines the frequency of pulsed laser mitigating peaks of melted pool in longitudinal direction (Point V_{P14} to V_{P24}, or V_{P25} to V_{P15} in Figure 10b), *x* range was defined between Δ_2 and $2\Delta_2$ (in Figure 8) in the experiment, so the scanning speed v_2 of pulsed laser was in the range of $2v_1 \leq v_2 \leq 2.5v_1$. Table 3 was the dual-beam laser polishing parameters.

Table 3. Parameters of the dual-beam laser polishing.

Set Up Laser Parameters	CW Laser	Pulsed Laser
Power	600 W	80 W
Wavelength	1080 nm	1064 nm
Beam profile pattern	Top-hat (shaped)	Top-hat
Pulse duration	N/A	1.3 μs
Spot diameter	0.47 mm	0.32 mm
Scanning speed	800 mm/s	2000 mm/s
Step-over	0.1 mm	0.1 mm
Scanning route	Zigzag	Zigzag-square wave
Top-hat beam profile	Between Medium and Ideal	

4. Results and Discussion

The experiment was taken according to the processes parameters given in Table 3. The white light interferometer (BRUKER WYKO Contour GT-K) was used again to measure the roughness of polished semisphere after polishing. Figure 11 is the measurement result, in which the sphere feature was removed to fit the true surface feature of roughness of *Sa* (=142 nm). Compared to Figure 5, the *Sz* value dropped much from 45931 to 27994 nm.

Figure 12 shows the actual polished semisphere associated with polishing performance data. The dual-beam took 35.5 s only to polish the semisphere surface resulted in the roughness reduction from initial state value of *Sa* (=877 nm) to post-polished value of *Sa* (=142 nm) as well as high laser polishing efficiency of 2890 cm²/H.

Numerous research have focused on the technology development of either CW laser polishing or pulsed laser polishing, very few studies have explored an approach to improve surface quality by the combination of CW laser polishing and pulsed laser polishing due to the large amount of the investment of such study and high uncertainty of experimental results.

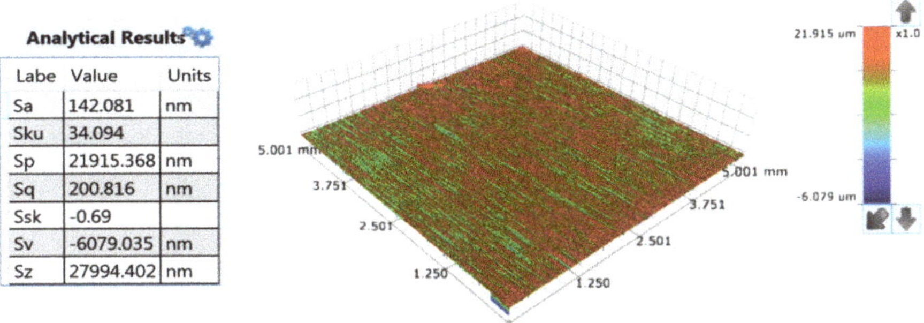

Labe	Value	Units
Sa	142.081	nm
Sku	34.094	
Sp	21915.368	nm
Sq	200.816	nm
Ssk	-0.69	
Sv	-6079.035	nm
Sz	27994.402	nm

Figure 11. Measurement result of the semisphere.

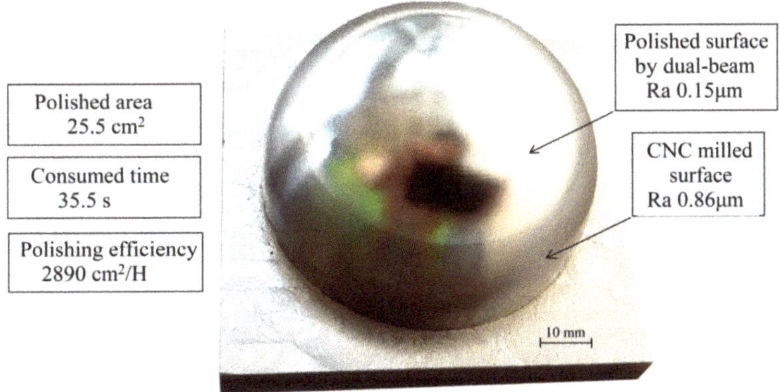

Figure 12. Polished semisphere by dual-beam.

The dual-beam laser polishing has the dual characteristics of both CW laser polishing and pulsed laser polishing, implementing the MPFV method effectively and leading to a decrease in peak-to-valley heights of the initial surface asperities. To achieve this, first of all, both individual laser processes are needed to be optimized, including setup of a series of appropriate parameters. In the experiment of CW laser polishing, the polishing parameters are given in Table 2, which had been optimized in previous works. A reverse beam extender and beam shaper was developed to make the spot irradiation of CW laser adjusted, ensuring that an appropriate spot diameter resulted in an optimized spot irradiation can be given under the condition that other parameters including top-hat beam profile of CW laser are fixed in advance.

The final surface quality achieved by PLμP relies on the performance of CW laser polishing. No matter how the PLμP is optimized, the final surface quality will not be improved if the surface roughness *Sa* achieved by CW laser polishing is more than 500 nm due to the evaporation of melted pool. To verify the importance of the optimization of CW laser polishing parameters, an experiment was conducted again using Table 2's parameters. The scanning speed was changed from 800 to 100 mm/s, and other parameters remained unchanged. It was found that the CW laser polishing resulted in a large area of molten pool evaporation on the top of the semisphere as shown in Figure 13. Obviously, the result would become worse if the high irradiation were used or "poor" top-hat profile were applied.

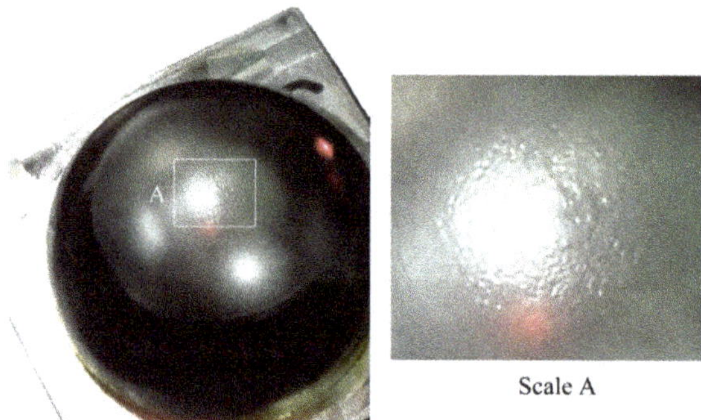

Scale A

Figure 13. Vaporization occurred on the top of semisphere.

5. Conclusions

The aim of this study was to explore an effective "melting peak for filling with valley" (MPFV) method to have dual-beam lasers polish steel freeform surface rapidly with smooth surface finish. The dual-beam consists of both a CW laser beam and a pulsed laser beam, whose profile patterns are top-hat. An initial rough surface was polished by a CW laser down to Sa (=426 nm) with high scanning speed 800 mm/s, meanwhile a pulsed laser plays a significant role to implement the MPFV method to polish the surface down to Sa (=142 nm), following up with the CW laser closely in the trajectory of zigzag-square wave. It took only around 35.5 s to polish a semisphere (ø60 mm) resulted in roughness reduction from initial state value of Sa (=877 nm) to post-polished value of Sa (=142 nm) of smoother surface finish. The dual-beam laser polishing efficiency is over 2890 cm^2/H, more than 100 times that of manual polishing process. The main conclusion to be drawn from this study is that the trajectory of zigzag-square makes the pulsed laser which is following up with CW laser implement MPFV method more effectively and efficiently than conventional PLμP process. Future research will attempt to investigate further the pulsed laser mechanisms associated with the adjustment of spot diameters as well as to determine the optimized polishing parameters that will ensure the higher polishing efficiency with smoother surface finish.

Author Contributions: Conceptualization, Y.Z. and Z.Z.; Data curation, H.X. and X.X.; Formal analysis, Y.Z. and H.X.; Funding acquisition, W.Z. and X. X; Methodology, Y.Z.; Z.Z. and W.Z.; Project administration, Y.Z. and Z.Z.; Resources, W.Z.; Software, H.X.; Validation, W.Z.; Visualization, H.X. and X.X.; Writing–original draft, Y.Z.

Funding: This research was funded by Shenzhen Science and Technology Plan (No. GGFW2017041209483817, No. JCYJ20170817112440533 and No. JCYJ20170817114441260).

Acknowledgments: The authors acknowledge all experimental engineers for their contribution of laser polishing experiments. In particular, Huayi Gong and Pingjia Li and Gearay team.

Conflicts of Interest: The authors declare no conflict of interest. The funders had no role in the design of the study; in the collection, analyses, or interpretation of data; in the writing of the manuscript, or in the decision to publish the results.

References

1. Burzic, B.; Hofele, M.; Mürdter, S.; Riegel, H. Laser polishing of ground aluminum surfaces with high energy continuous wave laser. *J. Laser Appl.* **2017**, *29*, 011701. [CrossRef]
2. Perry, T.L.; Werschmoeller, D.; Duffie, N.A.; Li, X.; Pfefferkorn, F.E. Examination of selective pulsed laser micropolishing on microfabricated nickel samples using spatial frequency analysis. *J. Manuf. Sci. Eng.* **2009**, *131*, 021002. [CrossRef]
3. Miller, J.D.; Tutunea-Fatan, O.R.; Bordatchev, E.V. Experimental analysis of laser and scanner control parameters during laser polishing of h13 steel. *Procedia Manuf.* **2017**, *10*, 720–729. [CrossRef]

4. Athanasiou, C.-E.; Bellouard, Y. A monolithic micro-tensile tester for investigating silicon dioxide polymorph micromechanics, fabricated and operated using a femtosecond laser. *Micromachines* **2015**, *6*, 1365–1386. [CrossRef]

5. Drs, J.; Kishi, T.; Bellouard, Y. Laser-assisted morphing of complex three dimensional objects. *Opt. Express* **2015**, *23*, 17355–17366. [CrossRef]

6. Bordatchev, E.V.; Hafiz, A.M.; Tutunea-Fatan, O.R. Performance of laser polishing in finishing of metallic surfaces. *Int. J. Adv. Manuf. Technol.* **2014**, *73*, 35–52. [CrossRef]

7. Goffin, N.; Tyrer, J.; Woolley, E. Complex beam profiles for laser annealing of thin-film cdte photovoltaics. *J. Laser Appl.* **2018**, *30*, 042006. [CrossRef]

8. Hafiz, A.M.K.; Bordatchev, E.V.; Tutunea-Fatan, R.O. Influence of overlap between the laser beam tracks on surface quality in laser polishing of AISI H13 tool steel. *J. Manuf. Process.* **2012**, *14*, 425–434. [CrossRef]

9. Bhaduri, D.; Penchev, P.; Batal, A.; Dimov, S.; Soo, S.L.; Sten, S.; Harrysson, U.; Zhang, Z.; Dong, H. Laser polishing of 3D printed mesoscale components. *Appl. Surf. Sci.* **2017**, *405*, 29–46. [CrossRef]

10. Beam Shaper/Top-Hat. Available online: https://www.holoor.co.il (accessed on 9 May 2019).

11. Kumstel, J.; Kirsch, B. Polishing titanium-and nickel-based alloys using cw-laser radiation. *Phys. Procedia* **2013**, *41*, 362–371. [CrossRef]

12. Chang, C.-S.; Chen, T.-H.; Li, T.-C.; Lin, S.-L.; Liu, S.-H.; Lin, J.-F. Influence of laser beam fluence on surface quality, microstructure, mechanical properties, and tribological results for laser polishing of SKD61 tool steel. *J. Mater. Process. Technol.* **2016**, *229*, 22–35. [CrossRef]

13. Vaithilingam, J.; Goodridge, R.D.; Hague, R.J.; Christie, S.D.; Edmondson, S. The effect of laser remelting on the surface chemistry of Ti6Al4v components fabricated by selective laser melting. *J. Mater. Proc. Technol.* **2016**, *232*, 1–8. [CrossRef]

14. Richter, B.; Blanke, N.; Werner, C.; Vollertsen, F.; Pfefferkorn, F. Effect of initial surface features on laser polishing of Co-Cr-Mo alloy made by powder-bed fusion. *JOM* **2019**, *71*, 912–919. [CrossRef]

15. Nüsser, C.; Sändker, H.; Willenborg, E. Pulsed laser micro polishing of metals using dual-beam technology. *Phys. Procedia* **2013**, *41*, 346–355. [CrossRef]

16. Ukar, E.; Lamikiz, A.; Martínez, S.; Tabernero, I.; de Lacalle, L.L. Roughness prediction on laser polished surfaces. *J. Mater. Process. Technol.* **2012**, *212*, 1305–1313. [CrossRef]

17. Pfefferkorn, F.E.; Duffie, N.A.; Morrow, J.D.; Wang, Q. Effect of beam diameter on pulsed laser polishing of S7 tool steel. *CIRP Ann.* **2014**, *63*, 237–240. [CrossRef]

18. Liu, R.; Wang, Z.; Zhang, Y.; Sparks, T.; Liou, F. A smooth toolpath generation method for laser metal deposition. In Proceedings of the 27th Annual International Solid Freeform Fabrication Symposium, Austin, TX, USA, 8–10 August 2016; pp. 1038–1046.

19. Roghanian, E.; Shakeri Kebria, Z. The combination of topsis method and Dijkstra's algorithm in multi-attribute routing. *Sci. Iran.* **2017**, *24*, 2540–2549. [CrossRef]

20. Rodríguez-Puente, R.; Lazo-Cortés, M.S. Algorithm for shortest path search in geographic information systems by using reduced graphs. *SpringerPlus* **2013**, *2*, 291. [CrossRef]

21. Ahmadi, S.M.; Kebriaei, H.; Moradi, H. Constrained coverage path planning: Evolutionary and classical approaches. *Robotica* **2018**, *36*, 904–924. [CrossRef]

22. Kapil, S.; Joshi, P.; Yagani, H.V.; Rana, D.; Kulkarni, P.M.; Kumar, R.; Karunakaran, K. Optimal space filling for additive manufacturing. *Rapid Prototyp. J.* **2016**, *22*, 660–675. [CrossRef]

23. Selvaraj, P.; Radhakrishnan, P. Algorithm for pocket milling using zig-zag tool path. *Def. Sci. J.* **2006**, *56*, 117–127. [CrossRef]

 © 2019 by the authors. Licensee MDPI, Basel, Switzerland. This article is an open access article distributed under the terms and conditions of the Creative Commons Attribution (CC BY) license (http://creativecommons.org/licenses/by/4.0/).

Article

Laser Surface Blasting of Granite Stones Using a Laser Scanning System

Joaquín Penide [1], Jesús del Val [1,*], Antonio Riveiro [1], Ramón Soto [1], Rafael Comesaña [2], Félix Quintero [1], Mohamed Boutinguiza [1], Fernando Lusquiños [1] and Juan Pou [1]

[1] Department of Applied Physics, University of Vigo, EEI, Lagoas-Marcosende, 36310 Vigo, Spain; jpenide@uvigo.es (J.Penide); ariveiro@uvigo.es (A.R.); rfsoto@uvigo.es (R.S.); fquintero@uvigo.es (F.Q.); mohamed@uvigo.es (M.B.); flusqui@uvigo.es (F.L.); jpou@uvigo.es (J.Pou)

[2] Department of Materials Engineering, Applied Mechanics and Construction, University of Vigo, EEI, Lagoas-Marcosende, 36310 Vigo, Spain; racomesana@uvigo.es

* Correspondence: jesusdv@uvigo.es

Received: 3 December 2018; Accepted: 18 February 2019; Published: 19 February 2019

Abstract: Granite stones are the most abundant rock of the crust. Due to their beauty, durability, and virtually zero maintenance, they have been used widely since ancient times in all types of construction, as a structural or decorative element. Commonly, this material is used with a polished finishing, but there has been an increased interest in giving it a rustic aspect, mainly for decorative or functional reasons, e.g., to reduce slipping. Rough surfaces are usually produced by means of bush hammering, but this is an extremely noisy and inefficient process. In this work we have explored the capabilities and limits of a laser blasting process assisted by a scanning system in order to produce precise and controllable roughness on two varieties of granite plates. It was found that laser blasting of thin granite tiles can be accomplished with processing widths up to 250 mm at medium-low laser power, obtaining a rustic aspect suitable for use in façades, paving, or flooring. Moreover, laser scanner systems are capable of enhancing the productivity of this process up to ten times greater than that found in previous works.

Keywords: surface treatment; CO_2 laser; scanning system; granite stone

---

1. Introduction

Granite stones are one of the most suitable materials for construction due to their beauty, durability, and low maintenance. This is a manifestly crystalline rock with an interlocking structure. Its composition depends on the exact variety of granite, but in general it is mainly composed of feldspar, quartz, a small amount of mica, and minor accessory minerals (such as zircon, apatite, magnetite, ilmenite, or sphene) [1]. These materials have a greater strength than marble, sandstone, or limestone and therefore are more difficult to machine [2].

The international trade of granite was valued at around \$2.5 billion in 2016 [3], with an estimated production of 20 million tons. From a commercial point of view, the maximum usage of granite is in paving, internal and external flooring, and façades [4].

Surface finish is one of the major concerns related to the application of natural stones. Different finishes will provide different aesthetic value, durability, or slip resistance. A polished finish is commonly used for internal flooring and tiling, while on the other hand, a rough finish, e.g., a bush hammered finish, is used to create a non-slip surface, ideal for high traffic outdoor areas or to give a rustic aspect employed on feature cladding and cobblestones.

Bush hammering is the most employed technique to give a rough surface finish to granites. It consists of beating the stone using a hammer designed specifically for this application, so some material is removed from the surface. Despite its wide application, it possesses some disadvantages:

The level of noise is high (>120 dB), the powder generation is pronounced, the minimum thickness to apply this method is restricted to high values (typically higher than 25 mm), and the cost of the process is increased by the replacement of worn-out tools.

In order to overcome these drawbacks, laser technology seems to be a good choice since it has already demonstrated its capability to treat different building materials such as concrete, natural stones, tiles, and rocks [5]. For example, some authors have focused on surface texturing of concrete by laser scabbling [6–8] or in marble [9]. Regarding granite, a polished surface was laser treated to be transformed into a non-slip surface by generating micro-craters [10,11] on it. The laser blasting of granite has been previously reported [12,13], which consisted of material being removed from its surface by a laser beam. Moreover, no mechanical stresses are induced on the plate, so thin tiles can be treated and wasted powder and noise is greatly reduced compared to conventional bush hammering.

Infrared lasers are typically employed for processing granite. The most common are Nd:YAG, diode, and CO_2 lasers, but CO lasers are also used [14]. A Nd:YAG laser is commonly utilized for surface treatments [15] and for producing craters [10]. On the other hand, drilling [16] and surface treatment [12,13] are also carried out by a diode laser. However, CO_2 lasers are most commonly employed for processing granite. Valente et al. [16] demonstrated that a CO_2 laser is more efficient than diode lasers for drilling white granite and it was assumed to be due to a higher absorption. Laser cutting of granite by CO_2 laser was recently reported in [17]. Drilling of granite submerged in water [18] and surface treatment [9] has also been carried out by CO_2 lasers.

The main limitation of laser technology for surface treatments (and particularly for laser blasting) is related to their difficulty in treating large areas due to the reduced dimensions of the spot size of a focused laser beam. However, the combination of high-power laser sources and scanning systems used in laser treatments, such as laser hardening [19], microstructuring [10], and more recently in laser cladding [20], can overcome the low productivity of laser blasting process.

In the present work, we investigated laser blasting process assisted by scanning systems on two varieties of granite. Therefore, the aim of this work was to present an in-depth study of the influence of the processing parameters (laser power, P, transverse speed (velocity in the Y-direction, v_1) and scanning speed (velocity in the X-direction, v_2)) on the roughness and quality of the finished surface for non-slip applications, as determined by advanced statistical methods.

2. Materials and Methods

2.1. Materials

The materials used in the experiments were two different granite plates: A white granite (called White Alba granite) and a black granite (a gabbro called Black South Africa granite). Both granites have a medium size grain, are compact, and exhibit an irregular fracture.

Plates of 350 mm × 250 mm × 10 mm and an initial average surface roughness of $R_a = 0.22$ μm and $R_a = 0.068$ μm for the white and black granites, respectively, were used in the laser blasting experiments.

2.2. Experimental Procedures

The laser treatment was performed using a 3.5 kW CO_2 slab laser (Rofin-Sinar DC035, Hamburg, Germany), emitting a TEM00 CW laser beam at a 10,600 nm wavelength. The laser beam was focused onto the surface of the sample using a lens with a 710 mm focal length, thus achieving a laser focal spot diameter of 0.56 mm. In order to perform a wide surface treatment, the laser beam was scanned over the sample, in all cases, in air at atmospheric pressure.

Figure 1 shows an illustration of the experimental set-up employed for the laser treatments. To perform a fast and wide surface treatment, a special designed polygon mirror scanner was implemented to produce a high-speed deflection of the laser beam in one direction (e.g., X-direction). On the other hand, the scanned laser beam was moved along a perpendicular direction (e.g., Y-direction) by means of a computer numerically controlled table. Using this scanning system, the laser beam was deflected up to 450 mm in the X-direction along the sample, therefore, tiles were processed in only one pass along the whole X-direction, while the scanned laser beam was moved at a constant rate in the Y-direction.

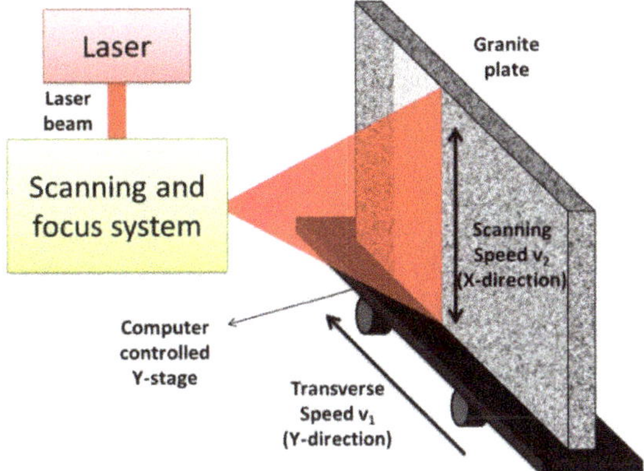

Figure 1. Illustration of the experimental set-up.

A full factorial design (FFD) approach was developed to screen out the key variables (laser power, P; transverse speed (velocity in the Y-direction, v_1); and scanning speed (velocity in the X-direction, v_2) which significantly influence the response variables (average roughness of samples, R_a and a quality factor, Q). Two levels (designated "+" and "−") for each of the four processing parameters were investigated, as summarized in Table 1. The processing parameters for the full factorial design were selected based on preliminary tests performed at the laboratory (not included here). Only factors with $p < 0.05$ were considered significant. In a full factorial experiment, responses were measured at all combinations (i.e., conditions at which the responses were measured) of the experimental factor levels. Using an analysis of variance (ANOVA) scheme, those factors having a significant effect on the response variables were identified. Additionally, a regression model was used to estimate the response values as a function of the significant variables. This experimental strategy allowed the determination of the influence of a reduced number of different laser parameter combinations without decreasing the accuracy of the results. Please consult [21] for more details about the mathematical procedure.

Table 1. Factors and levels for the 2^3 full factorial design.

Laser Power, P (W)		Transverse Speed, v_1 (mm/s)		Scanning Speed, v_2 (mm/s)	
−	+	−	+	−	+
1000	2500	0.83	3.33	100,000	200,000

2.3. Sample Characterization

Selected samples were inspected in frontal view to the laser treated area by means of an optical stereoscopic microscope (Nikon SMZ-10A, Tokyo, Japan) coupled to a photographic system in order to record and store images.

Roughness was measured by a surface roughness gauge (TESA Rugosurf 10G, Renens, Switzerland) in several locations of the treated areas. Then, an average value for the average roughness (R_a) was extracted to characterize the surface finishing after the laser treatment. Measurements were made in accordance with the recommendations specified by the International Standard ISO 4288:1996 [22].

In order to study the influence of laser radiation on the granite, laser treated and untreated samples of granite were analyzed by Raman spectroscopy. Raman reflection spectra were acquired by means of a spectrometer (Horiba Jobin Yvon LAbRam-HR800, Kyoto, Japan) equipped with an Ar laser excitation source (488 nm) and coupled with a microscope.

Finally, the quality of the laser blasted samples was evaluated by means of a quality factor (designated Q Factor) which takes into account different variables such as: melting of the surface; processing speed; laser power; amount of residue; removed material; roughness; and visual appearance.

Therefore, the following expression was used:

$$Q = X_0(20X_1 + 7.5X_2 + 7.5X_3 + 5X_4 + 5X_5 + 30X_6 + 25X_7) \tag{1}$$

The possible values for each variable are summarized in Table 2. The importance of each variable was weighted by a factor associated with its relevance on the final application, then, roughness, visual appearance, and detection of notorious melting of the surface were selected as the most relevant parameters determining the quality of samples for its application to produce non-slipping surfaces. On the other hand, the other variables are related with the laser process itself. Therefore, they are less important than the others, but still relevant.

Table 2. Possible values of those parameters involved in the definition of the quality of the laser blasted surfaces.

Parameters	Possible Values		
	High	Intermediate	Low
Blast (X_0)	1	0.5	0.1
Melting (X_1)	0.1	0.5	1
Processing speed (X_2)	1	0.5	0.1
Processing power (X_3)	1	0.5	0.1
Residues (X_4)	0.1	0.5	1
Removed material (X_5)	0.1	0.5	1
Roughness (X_6)	1	0.5	0.1
Appearance (X_7)	1	0.5	0.1

3. Results

3.1. Processing Results

The utilization of a laser beam in conjunction with a scanning system allows for the treatment of large areas. Processing results indicate that 250 mm in length (x-axis) can be treated in one pass using the experimental setup developed in this work. As seen in Figures 2 and 3, treated surfaces exhibit a homogenous aspect and a rustic aspect. The maximum productivity reached during the experimental tests was around 3 m^2/h for both granites. This processing rate is approximately ten times higher as compared to those results found in the literature [12,13]. It should be noted that the working rate using conventional methods, such as a rotary bush hammer, is around 15–20 m^2/h. In order to reach and surpass these figures, advantages of laser processing should be exploited, e.g., working in parallel with multiple laser beams or employing high brightness laser sources.

On the other hand, roughness values as high as 20 μm were reached during the processing of both granites. These values are smaller than in the case of conventional methods such as flaming (75 μm) or bush hammering (125 μm).

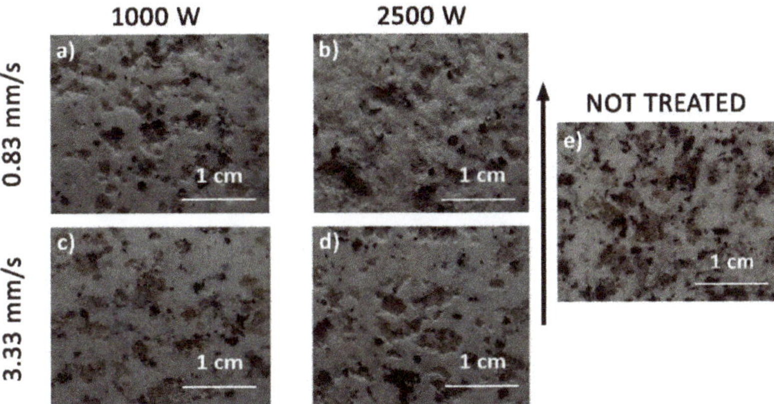

Figure 2. Images showing the aspect of the treated white granite as a function of the laser power and the transverse speed (**a–d**). The black arrow indicates the direction of the "fast" scan speed. (**e**) Image of not treated granite surface.

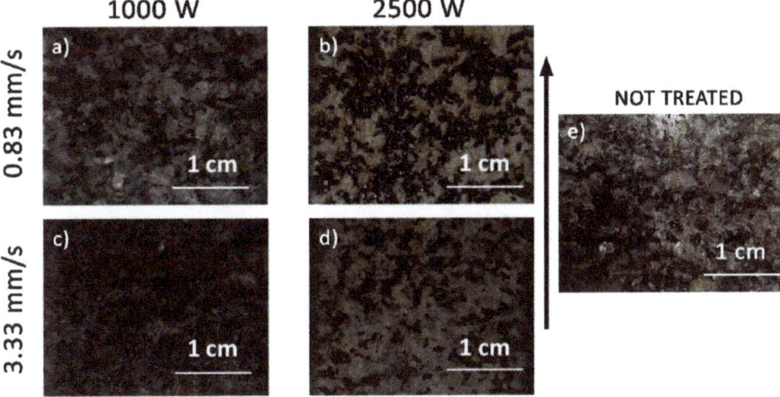

Figure 3. Images showing the aspect of the treated black granite as a function of the laser power and the transverse speed (**a–d**). The black arrow indicates the direction of the "fast" scan speed. (**e**) Image of not treated granite surface.

The interaction of a laser beam with rocks can spall, melt, or vaporize them as the energy transferred to the material raises its temperature locally. Literature data indicate that the most efficient rock removal mechanism is the thermal-spallation due to the lower temperature required as compared to melting or vaporization [23].

The scanning of a laser beam focused onto the surface of both granites, under an average irradiance of 10,960 W/mm^2, tends to produce the spallation of the surface, as depicted in Figure 4 for the black granite. This irradiance causes an increment in temperature which can be approximately calculated for a semi-infinite, homogeneous solid [24] by Equation (2):

$$T(x, y, z, t) = T_0 + \frac{Q}{2\pi k \sqrt{x^2 + y^2 + z^2}} \exp\left(\frac{-v\left(w + \sqrt{x^2 + y^2 + z^2}\right)}{2\alpha}\right) \tag{2}$$

where T—temperature; T_0—ambient temperature; Q—laser power; v—travel speed of the laser beam; $x/y/z$—distances in the x, y, and z directions; w—distance in the x direction in a moving coordinate of speed v ($w = x - vt$); α—thermal diffusivity; and k—thermal conductivity.

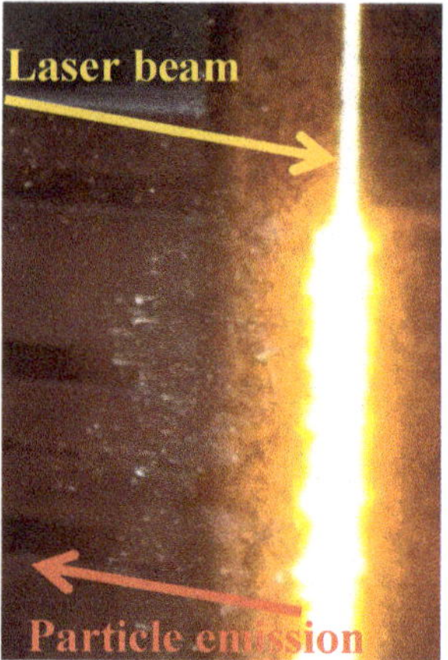

Figure 4. Emission of particles from the irradiated surface produced by thermal stresses induced in the black granite.

The increment in the temperature due to the laser scanning produces thermal stresses as given approximately by the well-known equation [25]:

$$\sigma_x = \sigma_y = \frac{EaT}{1-\nu} \tag{3}$$

where E—Young modulus; a—coefficient of lineal thermal expansion; T—temperature; and ν—Poisson's ratio.

As seen, thermal stress, and in consequence the spallation, is proportional to the temperature reached in the interaction region. These stresses are quite significant in granites due to their low thermal conductivity and high linear thermal expansions. The spalling/cracking is formed when the thermal stresses beneath the surface reach the tensile strength of the granite [10]. At this stage, flakes or particles are emitted from the surface in a normal direction as seen in Figure 4. It was observed that part of the flakes (up to 1–2 mm in length) can be melted and remain lightly adhered to the surface of the treated area during the processing at high laser powers and low scanning and traverse speeds as depicted in Figure 5.

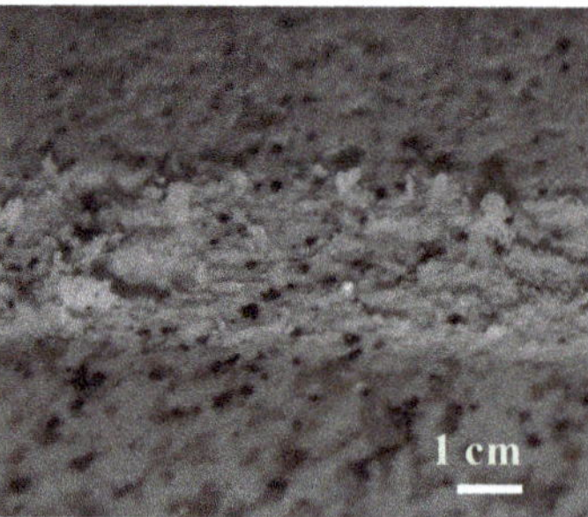

Figure 5. Flakes formed during the laser blasting of white granite remain adhered to the surface due to their melting.

During the processing of both granites, a slight melting of the surface can be observed. However, this effect is quite selective. Quartz and feldspar tends to be highly fractured, mainly in the white granite, however, mica tends to be essentially melted as corroborated by Raman analyses in processed samples of white granite (Figure 6). As noted, mica (annite) is in a glassy state.

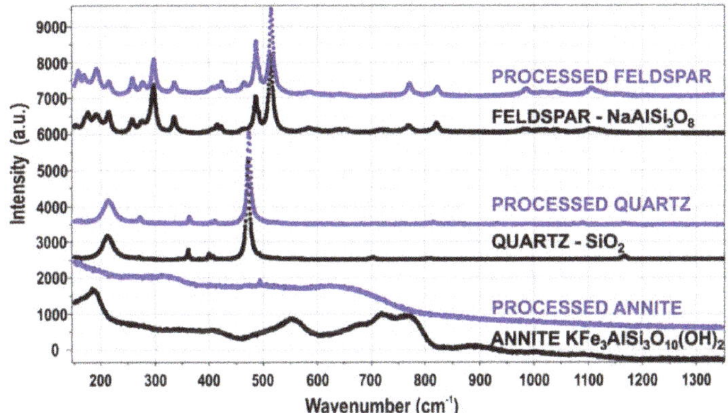

Figure 6. Raman spectrum for the different phases distinguished in the black granite prior and after the laser blasting process.

The melting process of mica is quite different in both granites. In the white granite, mica is only slightly melted and tends to form micrometric drops in order to minimize its surface energy (Figure 7a). On the other hand, mica particles in the black granite are massively melted and tend to form large drops, while feldspar and quart tends to bleach (Figure 7b).

(**a**) (**b**)

Figure 7. Localized melting observed in the (**a**) white and (**b**) black granites.

3.2. Roughness Results

As previously discussed, the appearance of the treated surfaces strongly depends on the processing parameters. The analysis of the full factorial design reveals laser power and transverse speed as the most statistically significant factors affecting the roughness during the processing of both granites. Influence of the scanning speed can be neglected for the range of values studied in this work. The response surfaces for the average roughness have been calculated as:

$$R_a(\mu m) = -10.72 + 0.0118P + 5.8375v_1 + 6.8 \times 10^{-5}v_2 \text{ (White granite)} \tag{4}$$

$$R_a(\mu m) = 11.00 + 0.0395P - 3.4152v_1 - 7.9 \times 10^{-5}v_2 \text{ (Black granite)} \tag{5}$$

Only terms up through Order 1 were included in the model for simplicity, and plotted in Figures 8 and 9.

As seen during the processing of both granites, the average roughness of the samples is increased by means of the increment of the laser power and by the reduction of the transverse and scanning speed. The effect of the reduction of the scanning speed is much less compared to the traverse speed. This can be explained under the basis of Equations (2) and (3). Utilization of high laser powers and low speeds produces large temperatures and, in consequence, high thermal stresses. The higher the thermal stresses, the more efficient the laser blasting process, and the sample exhibits a larger roughness.

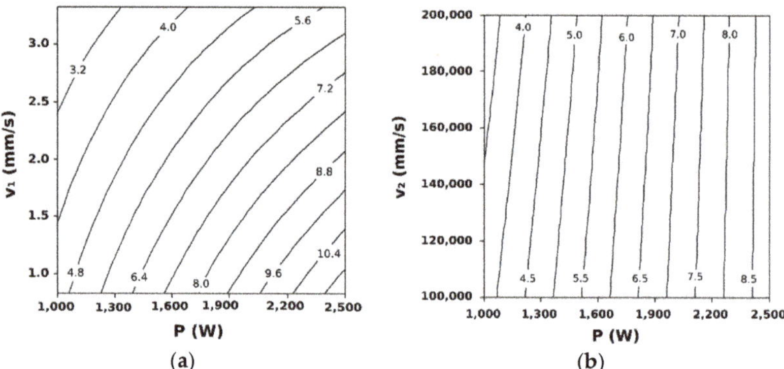

Figure 8. Contour plot of average roughness R_a (μm) versus (**a**) laser power (P) and transverse speed (v_1), and (**b**) laser power (P) and scanning speed (v_2) during the processing of white granite.

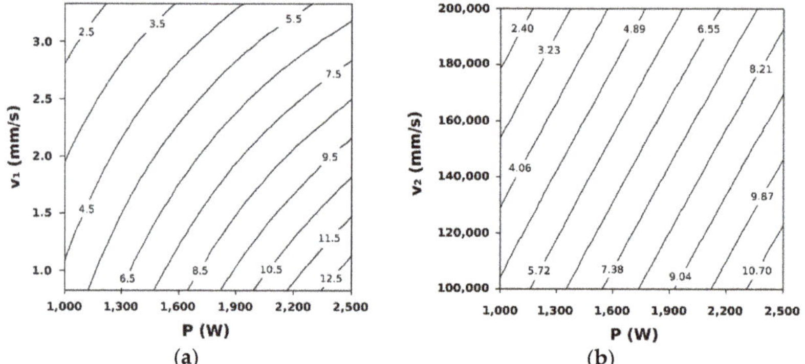

Figure 9. Contour plot of average roughness R_a (μm) versus (**a**) laser power (P) and transverse speed (v_1), and (**b**) laser power (P) and scanning speed (v_2) during the processing of black granite.

3.3. Quality Factor

A quality factor was proposed to evaluate the performance of the laser blasting process on the studied materials using the guidelines given in Section 2.3. The results of the full factorial design show that laser power and transverse speed are the statistically significant factors affecting the quality of the blasting. Influence of the scanning speed can be neglected for the range of values studied in this work. The response surfaces for the quality factor have been calculated as:

$$Q = 23.39 + 0.0168\,P - 11.11\,v_1 + 8.5 \times 10^{-6}\,v_2 \text{ (White granite)} \tag{6}$$

$$Q = 15.26 + 0.0177\,P - 6.16\,v_1 - 6.17 \times 10^{-5}\,v_2 \text{ (Black granite)} \tag{7}$$

Only terms up through Order 1 were included in the model for simplicity and plotted in Figure 10.

The quality obtained during the processing of the black granite is inferior to the quality obtained for the white granite. In both cases, the quality is increased by the increment of the laser power and the decrement of the transverse speed. Evolution of the quality as a function of the processing parameters is quite similar in both granites.

Laser blasting with the scanning system can be also employed in other kind of granites. Figure 11 shows a polished plate of Black Zimbabwe granite with three bands of non-slipping surface. They were made using 1000 W of laser power and 2 mm/s of transverse speed. Therefore, the CO_2 laser demonstrated its capability to produce a rustic non-slip surface on different kinds of granites.

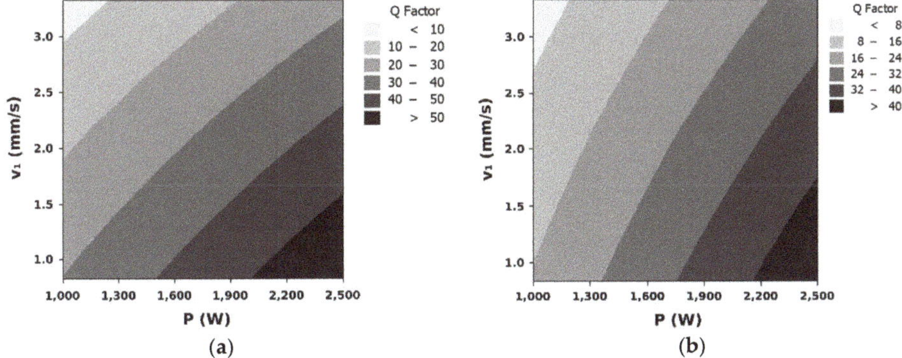

Figure 10. Quality factor evolution as a function of the laser power (P) and transverse speed (v_1) for the processing of the (**a**) white granite and (**b**) black granite.

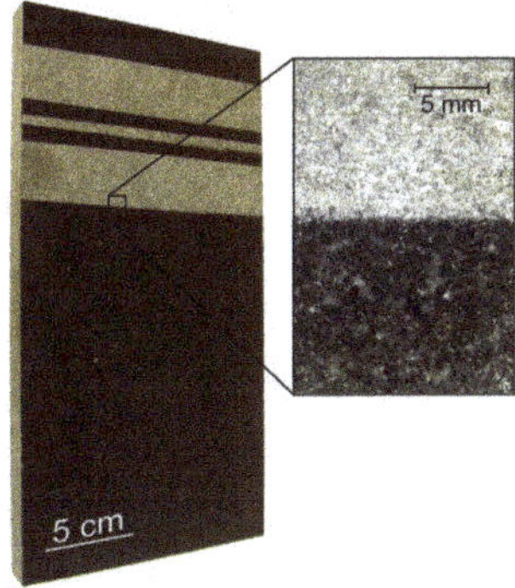

Figure 11. Image of a plate of Black Zimbabwe granite with three non-slip bands made by laser blasting and a detail of the transition from polished to laser treated surface.

4. Conclusions

A laser blasting process assisted by a scanning system has been used to treat the surface of thin granite tiles (10 mm). Using a moderate laser power (1 kW) this system allows blasting the surface of granite plates with processing widths of 250 mm. Processing conditions required to tailor the surface roughness and quality of results were determined using statistical methods. The quality of the laser blasted samples was evaluated by means of a quality factor (designated Q Factor) which takes into account different variables such as: melting of the surface; processing speed; laser power; amount of residue; removed material; roughness; and visual appearance. Laser power and traverse speed were found to be the most relevant processing parameters. Surface quality and roughness are both increased with the laser power and decreased with the traverse speed.

The obtained results show a way to maximize roughness and quality and reveal that the efficiency of the process is up to ten times higher than in prior works found in the literature using lasers.

Comparing to existing mechanical methods for granite blasting, such as bush hammering, the laser blasting process assisted by a scanning system here applied, shows several advantages: the level of noise is almost negligible compared with the 120 dB reached when using mechanical means; the laser blasting process allows blasting granites of just 10 mm, compared to the minimum thickness of 25 mm required to apply the bush hammering method; the amount of powder generated during the bush hammering is much higher than the one produced by the laser blasting technique.

Different kinds of granites (White Alba, Black South Africa and Black Zimbabwe) were successfully treated by means of a high-power CO_2 laser assisted by a scanning system, obtaining a rustic appearance surface suitable to be used in façades, paving, or flooring.

Author Contributions: Conceptualization, J.d.V. and J.P. (Juan Pou); Data Curation, J.d.V. and A.R.; Formal Analysis, A.R. and F.Q.; Investigation, J.P. (Joaquín Penide), J.d.V., R.S. and M.B.; Methodology, R.C.; Validation, F.L.; Writing—Original Draft Preparation, J.P. (Joaquín Penide); Writing—Review & Editing, A.R., F.L. and J.P. (Juan Pou).

Funding: This work was partially supported by the Xunta de Galicia (ED431B 2016/042, ED481D 2017/010, and ED481B 2016/047-0).

Acknowledgments: The authors wish to thank the technical staff from CACTI (University of Vigo) for their help with sample characterization and Marcos Nogueiras for technical assistance with the scanner system at early stages of the work. We would especially like to acknowledge Eduardo García from Cupastone for providing the granite samples and for his support along the whole work.

Conflicts of Interest: The authors declare no conflict of interest. The funders had no role in the design of the study; in the collection, analyses, or interpretation of data; in the writing of the manuscript, or in the decision to publish the results.

References

1. Blatt, H.; Tracy, R.; Owens, B. *Petrology: Igneous, Sedimentary, and Metamorphic*, 3rd ed.; W. H. Freeman: New York, NY, USA, 2005.

2. Goodman, R.E. *Introduction to Rock Mechanics*, 2nd ed.; Wiley: New York, NY, USA, 1989.

3. United Nations Comtrade Database. Available online: https://comtrade.un.org/ (accessed on 3 December 2018).

4. Taylor, H.A. *Compendium of World Dimension Stone Data*; Marble Institute of America: Oberlin, OH, USA, 2010.

5. Wignarajah, S. New horizons for high power lasers: Applications in civil engineering. In *High-Power Lasers in Civil Engineering and Architecture, Proceedings of International Forum on Advanced High Power Lasers & Applications (AHPLA), Osaka, Japan, 1–5 November 1999*; Nakai, S., Hackel, L.A., Solomon, W.C., Eds.; SPIE: Bellingham, WA, USA, 2000; pp. 34–44. [CrossRef]

6. Dowden, J.; Lazizi, A.; Johnston, E.; Nicolas, S. A thermoelastic analysis of the laser scabbling of concrete. *J. Laser Appl.* **2001**, *13*, 159–166. [CrossRef]

7. Lobo, L.M.; Williams, K.; Johnson, E.P.; Spencer, J.T. Particle size analysis of material removed during CO_2 laser scabbling of concrete for filtration design. *J. Laser Appl.* **2002**, *14*, 17–23. [CrossRef]

8. Peach, B.; Petkovski, M.; Blackburn, J.; Engelberg, D.L. An experimental investigation of laser scabbling of concrete. *Constr. Build. Mater.* **2015**, *89*, 76–89. [CrossRef]

9. Kosyrev, F.K.; Rodin, A.V. Laser destruction and treatment of rocks. In *ALT '01 International Conference on Advanced Laser Technologies, Proceedings of Advanced Laser Technologies (ALT '01), Constanta, Romania, 11–14 September 2001*; Dumitras, D.C., Dinescu, M., Konov, V.I., Eds.; SPIE: Bellingham, WA, USA, 2002; pp. 166–171. [CrossRef]

10. Hauptmann, J.; Wiedemann, G. Laser microstructuring of polished floor tiles. *Key Eng. Mater.* **2003**, *250*, 262–267. [CrossRef]

11. Panzner, M.; Lenk, A.; Wiedemann, G.; Hauptmann, J.; Weiss, H.J.; Ruemenapp, T.; Morgenthal, L.; Beyer, E. Laser processing of siliceous materials. In *High-Power Laser Ablation III, Proceedings of High-Power Laser Ablation, Santa Fe, NM, USA, 24–28 April 2000*; Phipps, C.R., Ed.; SPIE: Bellingham, WA, USA, 2000; pp. 621–634. [CrossRef]

12. Pou, J.; Trillo, C.; Soto, R.; Doval, A.F.; Boutinguiza, M.; Lusquiños, F.; Quintero, F.; Pérez-Amor, M. Laser blasting—A new method for surface treatment of dimension stones. *Key Eng. Mater.* **2003**, *250*, 247–252. [CrossRef]
13. Pou, J.; Trillo, C.; Soto, R.; Doval, A.F.; Boutinguiza, M.; Lusquiños, F.; Quintero, F.; Pérez-Amor, M. Surface treatment of granite by high power diode laser. *J. Laser Appl.* **2003**, *15*, 261–266. [CrossRef]
14. Bukatyj, V.I.; Perfil'ev, V.O. Effect of laser radiation on granite and marble. *Fiz. Khim. Obrab. Mater.* **2000**, *4*, 39–42.
15. Breaban, F.; Coutouly, J.F.; Braud, F.; Deprez, P. Nd:YAG laser beam-material interactions for marking and engraving: Application to alumina and granite. *Lasers Eng.* **2015**, *30*, 1–13.
16. Guedes Valente, L.C.; Pérez, M.A.A.; Gouvêa, P.M.P.; Martelli, C.; De Avillez, R.R.; Braga, A.M.B. Energy efficiency of drilling granite and travertine with a CO_2 laser and 980 nm diode laser. *Appl. Phys. Mater. Sci. Process.* **2013**, *110*, 639–642. [CrossRef]
17. Riveiro, A.; Mejías, A.; Soto, R.; Quintero, F.; Del Val, J.; Boutinguiza, M.; Lusquiños, F.; Pardo, J.; Pou, J. CO_2 laser cutting of natural Granite. *Opt. Laser Technol.* **2016**, *76*, 19–28. [CrossRef]
18. Kobayashi, T.; Nakamura, M.; Kubo, S.; Ichikawa, M. Drilling a hole in granite submerged in water by use of CO_2 laser. *J. Pet. Technol.* **2010**, *62*, 48–51. [CrossRef]
19. Lusquiños, F.; Conde, J.C.; Bonss, S.; Riveiro, A.; Quintero, F.; Comesaña, R.; Pou, J. Theoretical and experimental analysis of high power diode laser (HPDL) hardening of AISI 1045 steel. *Appl. Surf. Sci.* **2007**, *254*, 948–954. [CrossRef]
20. Nogueiras, M.; Penide, J.; Comesaña, R.; Del Val, J.; Quintero, F.; Boutinguiza, M.; Lusquiños, F.; Pou, J. Feasibility study of wide band laser surface treatment. *Phys. Procedia* **2013**, *41*, 356–361. [CrossRef]
21. Montgomery, D.C. *Design and Analysis of Experiments*, 8th ed.; Wiley: New York, NY, USA, 2012.
22. *ISO 4288:1996 Geometrical Product Specifications (GPS)—Surface Texture: Profile Method—Rules and Procedures for the Assessment of Surface Texture*; International Organization for Standardization: Geneva, Switzerland, 1996.
23. Xu, Z.; Reed, C.B.; Konercki, G.; Parker, R.A.; Gahan, B.C.; Batarseh, S.; Graves, R.M.; Figueroa, H.; Skinner, N. Specific energy for pulsed laser rock drilling. *J. Laser Appl.* **2003**, *15*, 25–30. [CrossRef]
24. Eagar, T.W.; Tsai, N.S. Temperature fields produced by traveling distributed heat sources. *Weld. J.* **1983**, *62*, 346–355.
25. Callister, W.D.; Rethwisch, D.G. *Materials Science and Engineering: An Introduction*, 8th ed.; Wiley: New York, NY, USA, 2011.

© 2019 by the authors. Licensee MDPI, Basel, Switzerland. This article is an open access article distributed under the terms and conditions of the Creative Commons Attribution (CC BY) license (http://creativecommons.org/licenses/by/4.0/).

Article

Microstructures and Wear Resistance of FeCoCrNi-Mo High Entropy Alloy/Diamond Composite Coatings by High Speed Laser Cladding

Haijiang Wang [1], Wei Zhang [1,*], Yingbo Peng [2,*], Mingyang Zhang [1], Shuyu Liu [1] and Yong Liu [1]

[1] Powder Metallurgy Research Institute, Central South University, Changsha 410083, China; wanghj@csu.edu.cn (H.W.); hugezmy123@gmail.com (M.Z.); Liu_SY@csu.edu.cn (S.L.); yonliu@csu.edu.cn (Y.L.)

[2] College of Engineering, Nanjing Agricultural University, Nanjing 210031, China

* Correspondence: waycsu@csu.edu.cn (W.Z.); ybpengnj@njau.edu.cn (Y.P.); Tel.: +86-731-88877669 (W.Z.)

Received: 14 February 2020; Accepted: 20 March 2020; Published: 24 March 2020

Abstract: FeCoCrNi-Mo high entropy alloy/diamond composite coatings were successfully prepared by high speed laser cladding. A high scanning speed was adopted (>30 mm/s), and the effects of laser power, scanning speed, and diamond content on the microstructure and wear resistance of the composite coating were studied. The processing parameters of laser cladding had significant influence on the dilution ratio, graphitization of diamond, and wear resistance of the composite coatings. When the laser cladding parameters were 3000 W of laser power and the high scanning speed of 50 mm/s, the composite coating exhibited a uniform microstructure, the lowest dilution ratio, and the best wear resistance. The wear resistance of the composite coating was enhanced with the addition of diamond, but microcracks also increased. When the amount of diamond was 15 wt.%, the best combination of microstructures and wear resistance was obtained.

Keywords: laser cladding; diamond composite coating; high entropy alloy; high scanning speed; wear resistance

---

1. Introduction

A diamond rope saw is a cutting tool with high hardness and flexibility, and which is widely used in stone cutting in mining and other industries [1–4]. Diamond string beads are the functional component in the diamond rope saw. Preparation methods of diamond string beads are mainly electroplating [5], sintering [6], and brazing methods [7]. The electroplating method is where the diamond is placed on the metal substrate to form a working layer, which is simple in technology but poor in wear resistance. The sintering method is to mix diamond and the matrix powders for granulating, and then combine them with the metal matrix for sintering. This method obtains good wear resistance and a long service life, but the process is complex, and the cost is high. For the brazing method, a layer of carbide is formed on the surface of the diamond by brazing filler metal containing strong carbide forming elements (Ti, Cr, W, etc.) to realize the chemical combination of diamond and matrix, so as to improve the wear resistance. However, in the brazing process, the retention of the matrix on the diamond abrasive particles is insignificant, and the abrasive particles easily fall off.

With the rapid development of laser technology in recent years, the laser cladding method has the advantages of preparing composite component materials with enhanced mechanical properties and fine and homogeneous microstructures [4]. In processing factors, laser cladding benefits from having unique features such as low heat affected zone, high cooling rate, and low dilution [8,9]. Laser cladding provides a novel idea for the preparation of diamond/metal matrix composite and its coating to avoid the shortcomings of the traditional method and achieve a high-performance workpiece [10–13].

Rommel [11], Iravani [12], and Mehrdad [14] used laser cladding technology to prepare diamond/metal matrix composite coatings. The interface bonding behavior between the diamond and metal matrix and the effect of laser cladding process parameters on microstructures and mechanical properties were studied. These results show that under the influence of different laser cladding process parameters, there are great differences in the thermal damage and graphitization of diamond particles, the bonding behavior between diamond and metal matrix, porosity and microcracks, and the dilution effect.

The combination of laser cladding technology and high entropy alloys is a new attempt in coating technology, and also a new method for the development of high entropy alloys (HEAs). HEAs have excellent high temperature stability, wear resistance, and corrosion resistance, so they exhibit great application potential in coating technology [15–17]. The rapid heating and cooling rate of laser cladding can refine the microstructures of the coating, and the combination of the two aspects can achieve the complementary effect and further increase the mechanical properties of the coating.

Moreover, the thermal stability of diamond is poor. In the laser cladding process of diamond–HEA composite coating, due to the concentration of laser energy, if the laser contacts the diamond for a long time, the tendency for diamond graphitization is intensified, and the HEA matrix and diamond can diffuse with each other to form carbides. Usually, for the metallic and alloy coating, the laser scanning speed is usually reduced (<20–30 mm/s) to obtain a coating with high bonding strength with the substrate. However, the thermal damage of diamond is irreversible.

Thus, high speed laser cladding was adopted in this study, which can not only make the coating have a good combination with the substrate, but also inhibits the graphitization of diamond and the carbonization reaction of metal components in HEA with diamond. Therefore, it is necessary to systematically study the microstructures and mechanical properties of the diamond composite coating by high speed laser cladding. In this study, FeCoCrNi-Mo HEA/diamond composite coatings were prepared by high speed laser cladding technology. The effects of processing parameters, such as laser power and scanning speed, and diamond addition on the microstructure, hardness, and wear resistance of the composite coatings was systematically discussed.

2. Experimental

Standard MBD4-type synthetic diamond (140–170 mesh average) was provided by Huanghe Whirlwind Co., Ltd. (Xuchang, China). Diamond was deposited in the magnetron sputtering coating equipment with argon atmosphere to form a Ti–Ni protective layer. Ti and Ni (inner Ti and outer Ni) with purity higher than 99.99% were used as targets. The layer was deposited about 15 µm thick. $Fe_{24.1}Co_{24.1}Cr_{24.1}Ni_{24.1}Mo_{3.6}$ (at.%) HEA powders prepared by gas atomization were applied as matrix materials, in which the average particle size was about 50 µm. The coated diamonds were mixed with HEA powders uniformly in the mixer for 5 h (mixer speed: 60 r/min); the coated diamond contents were 5 wt.%, 10 wt.%, 15 wt.%, and 20 wt.%, respectively.

FeCoCrNi-Mo HEA/diamond composite coatings were prepared on 42CrMo steel substrate by using laser cladding equipment with a laser line fiber-coupled semiconductor laser (Fraunhofer, Aachen, Germany), ABB manipulator (ABB, Zurich, Switzerland), and coaxial powder delivery. The laser powers were 3000, 4000, 5000, and 5500 W, and the scanning speeds were 30, 40, 50, and 60 mm/s, respectively. The laser spot diameter was 4.6 mm. The protective gas was argon with a 20 L/min flow rate. The surface of 42CrMo steel was polished, ultrasonic cleaned, and dried.

The hardness of the alloy was determined by using a Vickers hardness tester (BUEHLER5104, Buehler, Lake Bluff, American) under a 200 g load for 15 s and was averaged from three measurements. The wear properties were measured by HRS-2M high-speed reciprocating friction test equipment (HRS-2M, Zhongke Kaihua, Lanzhou, China) with a friction time of 15 min, 50 N loading, 15 Hz frequency (900 times/min), and a 5 mm stroke. A scanning electron microscope (SEM, FEI, Quanta 250 FEG, Vlastimila Pecha, Czech Republic) equipped with an energy dispersive spectrometer (EDS) was used to investigate the microstructure and chemical compositions of the samples.

3. Results and Discussions

3.1. Effect of Laser Power

3.1.1. Microstructures

Figure 1a shows the macro-morphology of the FeCoCrNi-Mo/diamond composite coating (diamond content 5 wt.%) prepared with different laser powers at high scanning speed of 50 mm/s. With the increase of laser power, the color of the coating surfaces changed from pale yellow to black. This shows that with the increase of laser power and the temperature of the molten pool, the laser energy density of the diamond particles increased, which led to the graphitization of the diamond particles. The C element of decomposed diamond reacted with the strong carbide forming elements in the HEA matrix, such as Cr, to form a large number of carbides. The typical microstructure of the coating by laser cladding is shown in Figure 1b. The coating can be divided into four areas, namely cladding layer, bonding area between substrate and coating, heat affected area, and substrate.

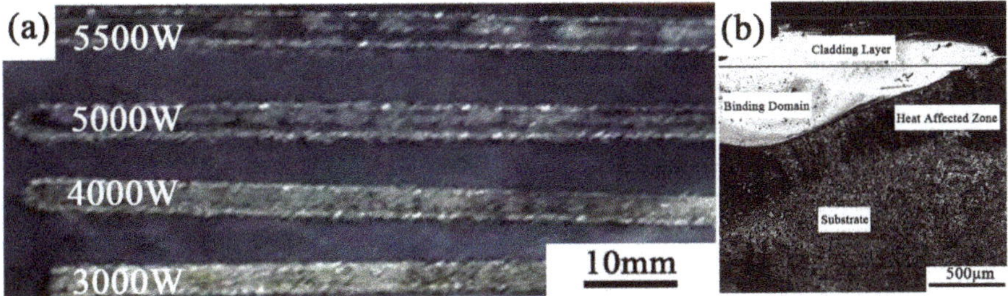

Figure 1. (**a**) Macro-morphology of FeCoCrNi-Mo/diamond composite coating prepared with different laser powers; (**b**) typical microstructure of coating (5000 W and 50 mm/s).

Figure 2 shows the longitudinal sectional morphology with a cross section inside of the coating under different laser powers. It can be seen that with the increase of laser power, the diamond content obviously decreased, and a lot of porosity appeared in the coating. When the power was 5500 W, obvious microcracks were found in the coating. The reason is that with the increase of laser power, the graphitization of diamond increased, and diamond particles began to decompose, which caused the declination of the diamond content. More free C atoms diffused into the HEA matrix, inevitably reacted with the oxygen introduced from the air, and formed a gas. The rapid solidification rate of laser cladding led to the gas not being able to be discharged and remained in the coating, forming micropores. On the other hand, the intensified diffusion of a C element resulted in the formation of a large number of carbides in the coating, which increased the residual stress in the coating and caused microcracks.

The dilution ratio of the composite coating prepared by different laser powers is shown in Figure 3a. The schematic diagram of coating and the calculation of dilution ratio η are shown in the Figure 3c and Formula (1), respectively. In Formula (1), h is the coating height and H is the depth of the bonding area.

$$\eta = \frac{h}{h + H} \tag{1}$$

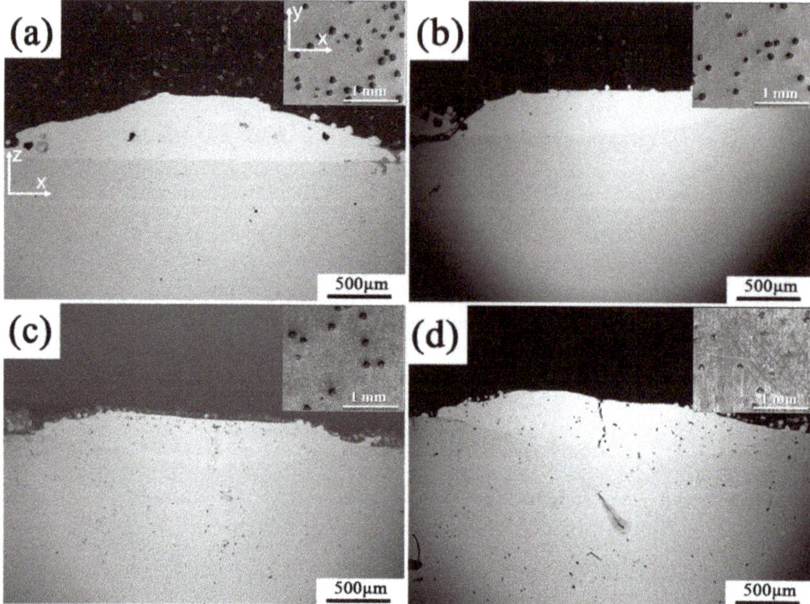

Figure 2. Microstructure of longitudinal section with cross section inside of FeCoCrNi-Mo/diamond composite coatings with different laser powers at a scanning speed of 50 mm/s; (**a**) 3000 W; (**b**) 4000 W; (**c**) 5000 W; (**d**) 5500 W.

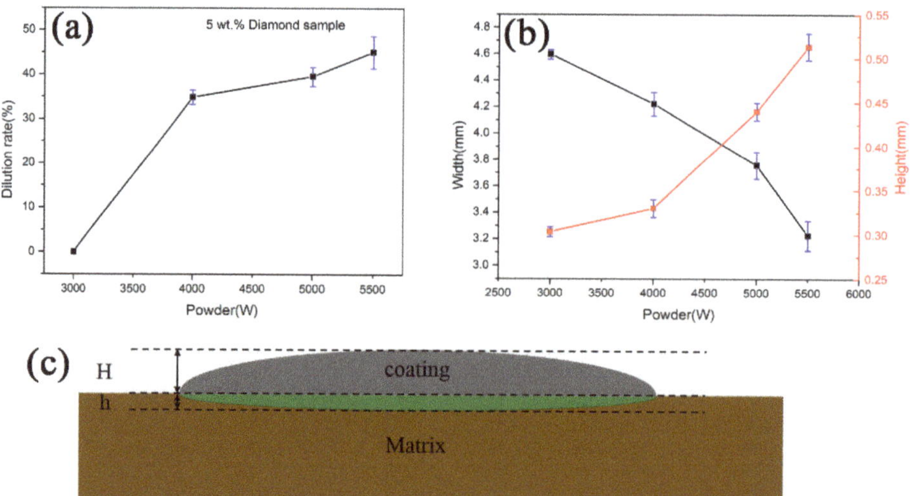

Figure 3. The variation curve of physical parameters of FeCoCrNi-Mo/diamond composite coating with laser power; (**a**) dilution ratio; (**b**) width and height; (**c**) the schematic diagram for the calculation of dilution ratio of coating.

The dilution ratio of coating refers to the change of coating composition caused by the diffusion of substrate elements into the coating due to the melting of the substrate during laser cladding. When the dilution ratio is too low, the bonding force between the substrate and the coating is reduced, and metallurgical bonding cannot be formed, and even the coating will fall off. When the dilution ratio is too high, the matrix will over-dilute the coating, which will increase the uncertainty of the coating composition, increase the tendency of cracking and deformation, and reduce the coating performance, as shown in Figure 3a, with the selected scanning speed of 50 mm/s, and the dilution rate of the coating increased with the increase of laser power. When the laser power was 3000 W, the dilution ratio of the coating was almost zero; when the laser power was increased to 4000 W, the dilution ratio of the coating increased sharply to 35%; when the laser power was further increased, the variation of the dilution ratio became gentle gradually, and finally reached 47% at 5500 W.

When the laser power was low, the temperature of the molten pool was relatively low, so the heat affecting the substrate was weak with a low dilution rate. On the other hand, when the laser power was high, the laser energy could melt the powders and effected the substrate, improving the metallurgical bonding. However, at the same time, too many elements of the substrate were introduced into the coating, which had an impact on the performance of the coating.

During laser cladding, there is splashing or ablation of powder under the laser condition. As a result, not all powders are utilized. In order to calculate the utilization ratio of powder, according to [18], the formula for the utilization ratio of powder is as follows:

$$\lambda = \frac{\rho B H V_s}{V_f} \tag{2}$$

In Formula (2), λ is the powder utilization rate (%), B and H are width and average height of the coatings (mm), respectively, V_s is scanning speed (mm/s), V_f is the powder delivery rate (g/s), and ρ is powder density (g/cm^3). The relationship between laser power and coating size is shown in Figure 3b when the scanning speed V_s and the powder feeding rate V_f were constant. According to Formula (1), when the laser power was 3000 W, the powder utilization rate was the largest, and it decreased slightly with the increase of laser power. This is because the increase of laser power made the thermal effect on the substrate, becoming stronger gradually, and the increase of the dilution ratio led to the decrease of the size of the coating.

3.1.2. Mechanical Properties

Figure 4a shows the microhardness of composite coatings with laser power. The trend of microhardness change with laser power was to increase first and then decrease. When the laser power was 3000 W, the average hardness of the coating was 329 HV$_{0.2}$; when the laser power was 5000 W, the maximum average microhardness of the coating reached about 404 HV$_{0.2}$. However, when the laser power continued to increase to 5500 W, the microhardness decreased to 333.28 HV$_{0.2}$. The reason is that at high laser power, the higher dilution ratio led to serious diffusion of substrate into the coating, resulting in the inhomogeneous composition and microcracks in the coating. These microcracks reduced the continuity of the coating and released the internal stress of the coating, resulting in the obvious reduction of the microhardness.

A friction test is the method to reflect the grinding or wear resistance of materials. The friction coefficient tends to be gentle and stable when the materials have good and stable wear resistance. Figure 4b shows the friction coefficient–time curves of the composite coatings prepared by different laser powers. When the scanning speed was 50 mm/s, with the increase of laser power, the friction coefficient of the coating decreased first and then increased. When the laser powers were 3000 W and 5500 W, the friction coefficient of the coatings was about 0.5, which was higher than the intermediate laser power of about 0.3. Therefore, the wear resistance of coatings were further determined by the worn surface of the counterpart Si$_3$N$_4$ ball. Figure 5 shows the worn surface of Si$_3$N$_4$ dual ball against the coatings prepared with different laser powers. Figure 5a–h corresponds to the laser power at 5500, 5000, 4000, and 3000 W, respectively, and the worn area of the corresponding Si$_3$N$_4$ ball was 3.08, 3.38, 3.14, and 12.82 mm^2, respectively. It can be seen from Figure 5a,b that the worn area of high power (5500 W) to the Si$_3$N$_4$ ball was the smallest, and the worn surface was relatively flat with few ploughings. With the decrease of laser power, the worn area increased, and the ploughing of the worn surface of Si_3N_4 ball increased, as shown in Figure 5c–f. At the lowest laser power (3000 W), the area of wear mark and the number of ploughing increased sharply, the wear mechanism changed from adhesive wear to abrasive wear, and the wear resistance of coating was the best, as shown in Figure 5g,h.

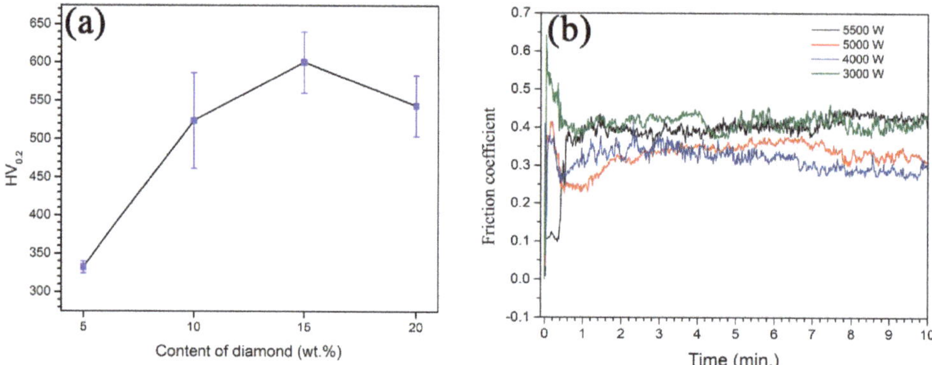

Figure 4. Mechanical properties of FeCoCrNi-Mo/diamond composite coatings prepared with different laser powers: (**a**) microhardness; (**b**) friction coefficient–time curves.

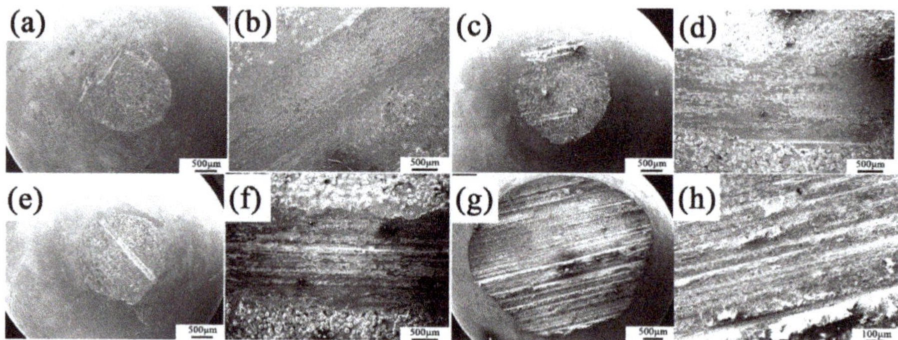

Figure 5. The wear morphology of FeCoCrNi-Mo/diamond composite coating was prepared by different laser powers; (**a,b**) 5500 W; (**c,d**) 5000 W; (**e,f**) 4000 W; (**g,h**) 3000 W.

The reason why the wear resistance of the coatings was worse with the increase of power is that the wear resistance of HEA/diamond composite coatings lies in the cutting effect of diamond particles retained by the HEA matrix. Therefore, whether the high hardness and wear resistance of diamond can be maintained after laser cladding is the key factor to determine the performance of coatings. However, the thermal stability of diamond is relatively poor. If the heating temperature of diamond is higher than 840–940 °C in air or 950–1100 °C in a vacuum, oxidation and graphitization occur [18]. In addition, a diamond is easy to react with some transition metal elements (most of them are strong carbide forming elements) to form carbides in the process of coating preparation, which also cause diamond failure.

Considering the microstructure and mechanical properties of FeCoCrNi-Mo/diamond composite coatings prepared by different laser powers, when the laser power was 3000 W, the coating was relatively optimal. Thus, the scanning speed is changed under the condition of 3000 W to explore the influence of scanning speed on the microstructures and properties of the composite coating.

3.2. Effect of Scanning Speed

3.2.1. Microstructures

The scanning speed also has an important influence on the microstructure and performance of the composite coating. Low scanning speed causes serious ablation of the powders; high scanning speed increases the amount of un-melted powders in the coating, resulting in poor bonding between the coating and the substrate. Therefore, FeCoCrNi-Mo/diamond composite coatings with the same 5 wt.% diamond addition were prepared by using the scanning speed of 30, 40, 50, and 60 mm/s with the specific laser power of 3000 W. It should be noted that the speed range of 30–60 mm/s is higher than normal laser cladding for metallic materials or alloys. The morphology of the coatings is shown in Figure 6.

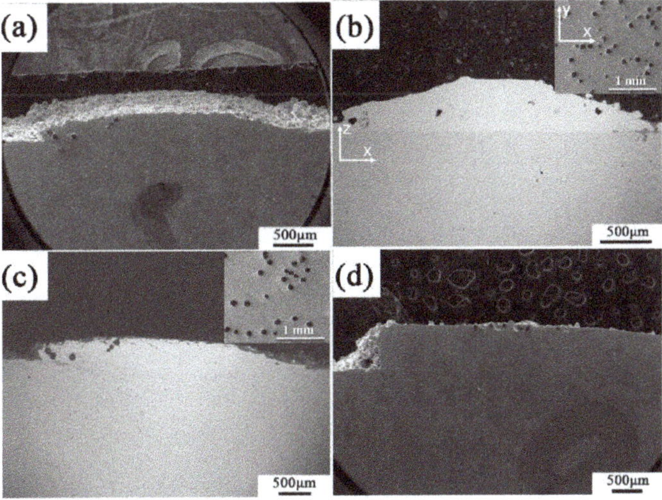

Figure 6. The microstructures of longitudinal section with cross section inside of FeCoCrNi-Mo/diamond composite coating was prepared at 3000 W laser power with different laser scanning speeds; (**a**) 60 mm/s; (**b**) 50 mm/s; (**c**) 40 mm/s; (**d**)30 mm/s.

It can be seen from Figure 6a that there were a lot of un-melted HEA powder particles on the coating surface due to the high scanning speed; un-melted powders can lead to poor bonding between the coating and substrate and result in micro-cracks. With the decrease of scanning speed, the width of the coating decreased, but the dilution ratio of the coating increased, as shown in Table 1. This is because the low scanning speed enhanced the interaction time between laser and powder/substrate, resulting in a larger heat affected zone of the coating; thus, the dilution ratio of the coating increased, accompanied with the declination of diamond content, as shown in Figure 6b,c. As the scanning speed further decreased to 30 mm/s, it can be seen from Figure 6d that the absorption rate of the coating for laser energy reached the highest, the dilution rate was the largest, and there was a lot of ablation of HEA powders. Moreover, long-term contraction between diamond particles and the laser at such a low scanning rate would inevitably lead to serious oxidation and graphitization of the diamond and loss by its own wear-resistant characteristics.

Table 1. The characteristics of FeCoCrNi-Mo/diamond composite coating prepared at different scanning speeds with laser power of 3000 W.

Scanning Speed (mm/s)	Width (mm)	Height (mm)	Dilution Ratio (%)
60	4.02	0.32	0
50	3.88	0.33	<10
40	2.77	0.34	28.24
30	2.31	0.33	43.58

3.2.2. Mechanical Properties

Due to the deterioration of coating properties caused by higher or lower scanning speed, the mechanical properties of coatings prepared by moderate scanning rate were investigated, such as 40 and 50 mm/s. When the scanning speed was 40 and 50 mm/s, the average micro-hardness of the coating was 329 $HV_{0.2}$ and 312 $HV_{0.2}$, respectively. The micro-hardness of the two coatings was close to each other, so the trends of friction coefficient–time curve and friction coefficient were similar. As shown in Figure 7a, when the scanning speed was 50 mm/s, the initial friction coefficient of the coating was higher than the 40 mm/s sample, and the curves turned to be stable quickly, slightly higher than the 40 mm/s sample, with the average friction coefficient of 0.41 and 0.37, respectively. Figure 7b,c shows the worn surface of the Si_3N_4 dual ball against the coatings prepared by the scanning speed of 40 and 50 mm/s, respectively. Compared with the sample by 50 mm/s (12.82 mm^2), the worn area of the dual ball was significantly reduced in the 40 mm/s sample to only 3.98 mm^2. It exhibited a mixed wear mechanism of adhesive wear and abrasive wear. When the scanning speed was reduced from 50 to 40 mm/s, due to the increasing contact time between diamond particles and laser, the graphitization of diamond became more serious, and the wear-resistance of the coating was significantly reduced.

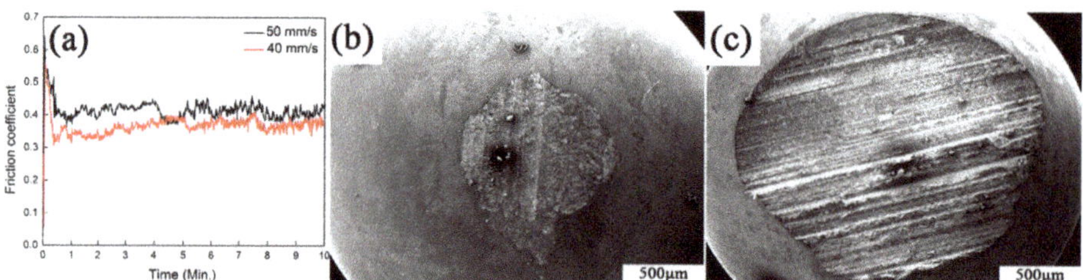

Figure 7. Frictional properties of FeCoCrNi-Mo/diamond composite coating. (**a**) The coefficient of friction; worn surface of dual ball by (**b**) 40 mm/s; (**c**) 50 mm/s.

3.3. Morphology of Diamond

The existence of the diamond in the composite coating has a significant effect on the retention of a matrix of diamond, and the grinding effect of the coating. It is expected that the graphitization of diamond and serious diffusion between diamond and HEA are restrained while the metallurgical bonding of diamond with HEA matrix is realized. Table 2 shows the grinding effect of composite coatings with different laser power and scanning speed, which is inextricably related to the morphology of diamond in the coating.

Table 2. The grinding effect of composite coatings with different laser power and scanning speed.

Laser Power (W)	Scanning Speed (mm/s)	Grinding Effect
5500	50	Bad
5000	50	Bad
4000	50	Bad
3000	50	Good
3000	40	Bad
3000	30	Bad
3000	60	Bad

During the laser cladding process, when the laser power was 3000 W and scanning speed was 50 mm/s, HEA powder completely melted and formed good bonding with diamond particles in the FeCoCrNi-Mo/diamond composite coating, as shown in Figure 8a. Meanwhile, there was strong metallurgical bonding between the HEA matrix and diamond through the diffusion of Cr, which played an important role in improving the retention of matrix to diamond. Only a small amount of C in the diamond changed into free C atoms, which formed metallurgical bonding with HEA matrix and partially diffused into the HEA matrix to realize the interstitial solution strengthening of the matrix and further improved the retention. These results were confirmed in our previous study [19]. On the other hand, the dilution ratio of the coating was relatively low; only a small number of elements of steel substrate diffused into the coating and the effect on the diamond was small under the condition of very fast cooling rate in laser cladding. The diamond particle could maintain the original crystal shape and its own good mechanical properties, so it had a good grinding effect, as shown in Figure 8a and Table 2.

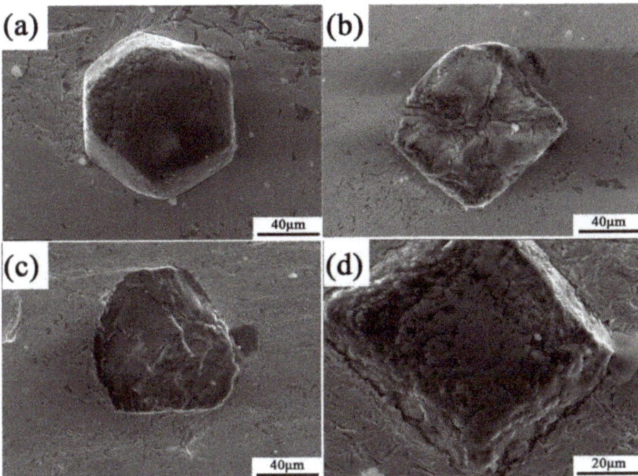

Figure 8. The morphology of diamond bonding in different laser energy density composite coating; (**a**) 3000 W and 50 mm/s; (**b**) 3000 W and 40 mm/s; (**c**) 5500 W and 50 mm/s; (**d**) 3000 W and 60 mm/s.

When the scanning speed decreased to 40 mm/s, with longer contact time between diamond and laser, under the synergistic effect of oxidation and graphitization, the diamond particle had an obvious structural transformation, gradually lost its crystal shape, the edges and corners became smooth and the size decreased, and the wear resistance decreased significantly, as shown in Figure 8b. As with the similar situation of diamond, when the laser power increased to 5500 W, as shown in Figure 8c, the diamond had obvious graphitization and completely lost its crystal shape, which meant that the diamond particle as an abrasive had failed. Although it could form a strong metallurgical bond with the HEA matrix, the diamond was seriously damaged and failed, resulting in the decrease of friction properties of the coating. In addition, when the scanning speed was too high (60 mm/s in Table 2), as shown in Figure 8d, the diamond could maintain a complete crystal shape and performance, but due to the shorter contact time between diamond and laser, there were a lot of un-melted powders in the HEA matrix, there was no metallurgical bonding with the diamond, and there were obvious gaps in the combined area of the diamond and the HEA matrix. In the process of friction, the retention of the HEA matrix of the diamond particles decreased, which would lead to the peeling off of the diamond, thus making the coating failure. Based on the morphology of the diamond and grinding effect of the coatings, the optimal processing parameters of FeCoCrNi-Mo/diamond composite coating by laser cladding were 3000 W and 50 mm/s.

3.4. Effect of Diamond Content

3.4.1. Microstructures

The diamond concentration (e.g., when the diamond volume fraction is 25%, the diamond concentration is 100%) is an important factor to determine the wear resistance of the diamond composite coating. Generally speaking, the higher the diamond concentration, the better the hardness and wear resistance of the coating. However, due to the poor compatibility between diamond and metal matrix, when the concentration is too high, the characteristics of the composite coating will be changed, resulting in poor bonding of the coating and the substrate, making the coating failure in service. Therefore, the appropriate amount of diamond was studied to improve the mechanical properties of the coating while maintaining a strong bonding strength with the substrate.

The optimal processing parameters of laser cladding were used to prepare the composite coatings with different diamond content. The diamond addition was 5 wt.%, 10 wt.%, 15 wt.%, and 20 wt.%, respectively. The cross-section microstructure of the coating with different diamond content is shown in Figure 9. The bonding of diamond and steel substrate was relatively good. The diamond particles were mainly distributed on the upper part of the coating. This is because the HEA matrix formed the molten pool during laser cladding; however, the diamond density was relatively low, and the diamond particles were suspended in the molten pool and retained these microstructures until room temperature. With the increase of the diamond addition, cracks began to appear at the edge of the coating, as shown in Figure 9c,d. As a result of the increasing diamond content, the wettability of the mixed powder and steel substrate became worse, and cracks were more likely to form in the coating.

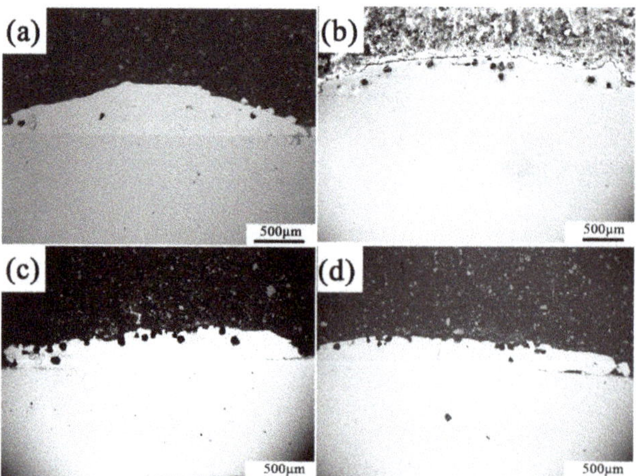

Figure 9. The longitudinal section morphologies of the composite coatings with different diamond content; (**a**) 5 wt.%; (**b**) 10 wt.%; (**c**) 15 wt.%; (**d**) 20 wt.%.

The microstructures of the HEA matrix with different diamond content are shown in Figure 10. When the content of diamond changed from 5 wt.% to 10 wt.%, the microstructure of the coating was obviously refined. When the diamond content continued to increase to 15 wt.% and 20 wt.%, the HEA matrix in the coating exhibited dendrite growth and micro-cracks.

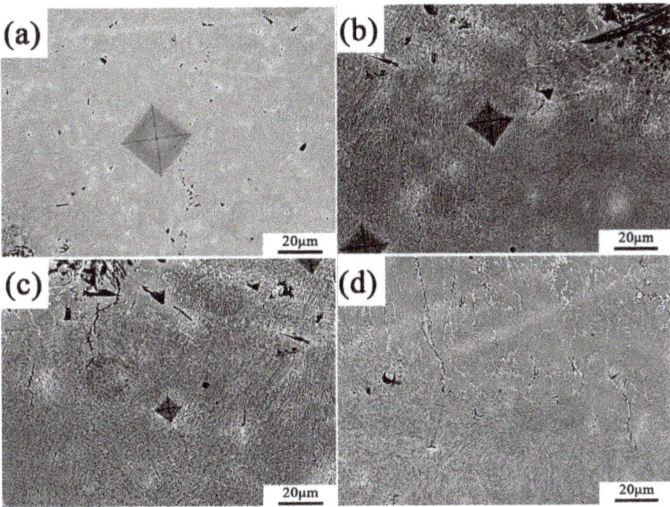

Figure 10. Microstructure of HEA matrix in composite coatings with different diamond content; (a) 5 wt.%; (b) 10 wt.%; (c) 15wt.%; (d) 20 wt.%.

The C element diffused into the matrix from the diamond is a typical kind of structure-refining element, which can refine the HEA structures significantly [20]. As a heterogeneous phase, the surface of diamond particles can be used as solid–liquid interface for secondary growth of the alloy during solidification. Therefore, the microstructures of HEA were refined obviously. On the other hand, due to the solid solution of the C element entering the HEA matrix and the addition of new growth interfaces of the diamond surface, the non-equilibrium effect of solidification was intensified. Therefore, cellular growth when the content of diamond was 5 wt.% changed to dendrite growth when the content of diamond was high.

The morphology of the interface between the diamond and HEA matrix is shown in Figure 11a. There were four phases around diamond particles, namely black phase, dark gray needle-like, and massive phase, as observed in Figure 11b,c, respectively, light gray matrix phase, and white veinlet phase. EDS results are listed in Table 3. C elements from diamond particles diffused into HEA and combined with Cr elements to form Cr rich carbide phase, which was richer in Cr and C (black phase). The composition of the light gray phase was closed to the theoretical composition of a FeCoCrNi-Mo HEA matrix with interstitial solid strengthening of C element. The dark grey phase exhibited a relatively high Mo content and less Cr, referred to as the intermetallic σ phase, because the Mo element can promote FCC (Face Center Cubic) σ phase transition of the FeCoCrNi system [21], and has been proved in our previous studies using FeCoCrNi-Mo HEA with the same composition as that in this study [22,23]. The white phase had the highest Mo content and a relatively high C content, which was the Mo-rich carbide phase.

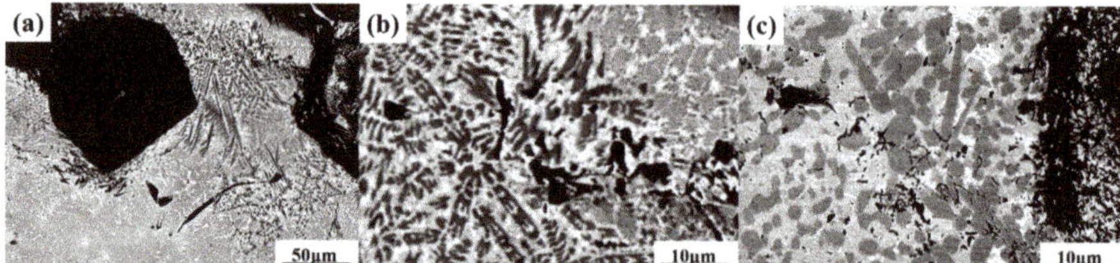

Figure 11. Microstructures of HEA/diamond coating. (**a**) The bonding area; (**b**) HEA matrix; (**c**) around diamond particles.

Table 3. Composition (at.%) of different phases (average of three measurements).

Phase	Fe	Co	Cr	Ni	Mo	C
Black phase	12.01	8.27	30.15	5.41	2.65	41.50
Light grey phase	19.53	18.50	22.48	23.90	4.65	21.34
Dark grey phase	17.71	19.58	5.70	28.84	6.88	23.36
White phase	10.23	17.92	8.96	11.65	9.87	39.90

Among all alloying elements in HEA, Cr and Mo have the largest enthalpy of mixing (more negative), which is −61 and −67, respectively [24]. Compared with other elements, they are easier to form carbides, as shown in Table 4.

Table 4. Mixing enthalpy between C and the alloying elements in HEA (kJ/mol) [24].

Element	Fe	Co	Cr	Ni	Mo
C	−50	−42	−61	−39	−67

3.4.2. Mechanical Properties

Figure 12a shows the microhardness of FeCoCrNi-Mo/diamond composite coatings with different diamond addition. It can be seen that the microhardness of the composite coating increased first and then decreased, with the increase of diamond content. The hardness was the highest when diamond content was 15 wt.%. This is because the increase of diamond content improved the overall hardness of the composite coating. The C element diffused into the coating and formed the hard second phase such as chromium carbides and molybdenum carbides. Meanwhile, C also played the role of interstitial solution strengthening of the HEA matrix. Therefore, the coating was strengthened from many aspects, and the hardness of the coating was gradually improved. However, when the diamond addition was 20 wt.%, there were many micro-cracks in the composite coating, which released part of the internal stress, resulting in the decrease of microhardness of the coating.

The friction coefficient–time curves of the diamond composite coatings are shown in Figure 12b. When the diamond content was 5 wt.%, the initial friction coefficient of the coating was high, and the friction coefficient remained high after stabilization. With the increase of diamond content, the friction coefficient decreased and became more stable, showing better wear resistance.

Figure 13 shows the worn surface of the Si_3N_4 dual ball. With the increase of diamond content, the coating was obviously elevated in grinding effect, and the worn surface area of the dual ball became larger. However, when the diamond was 20 wt.%, due to the appearance of cracks in the coating, the peeling of the coating occurred during the friction process, and the grinding performance of the coating decreased, so the worn area of the dual ball surface decreased. When the diamond content of the coating was 10 wt.%, 15 wt.% and 20 wt.%, the worn area of the dual ball was 15.48, 18.19 and 16.63 mm^2, respectively; compared with 12.82 mm^2 of 5 wt.%, coatings with higher diamond content had good wear resistance and grinding effect.

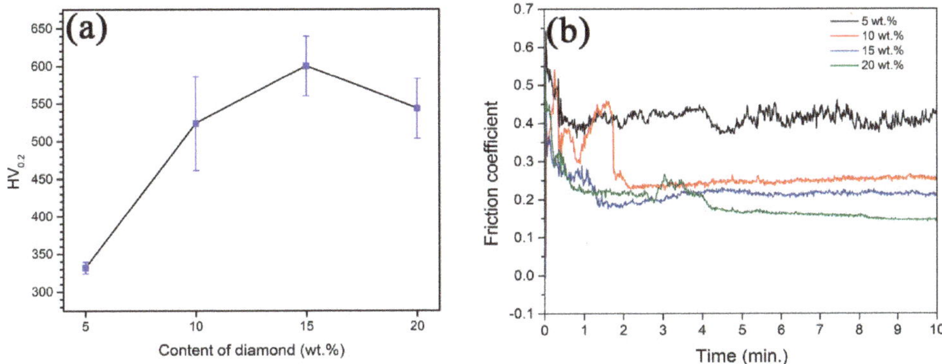

Figure 12. Mechanical properties of composite coatings with different diamond content. (**a**) Microhardness; (**b**) the coefficient of friction.

Figure 13. The wear morphology of the dual ball with different diamond content composite coating; (**a**) 10 wt.%; (**b**) 15 wt.%; (**c**) 20 wt.%.

4. Conclusions

- FeCoCrNi-Mo HEA/diamond composite coatings with good wear resistance were successfully prepared by high speed laser cladding technology.
- Micro-holes, cracks, and graphitization of diamond appeared and aggravated with the increase of laser power, resulting in the decrease of hardness and wear resistance. Too low a scanning speed increased the energy absorption of powder and the graphitization of diamond; too high a scanning speed resulted in the increase of un-melted HEA powders and poor bonding between HEA and the diamond.
- When the high scanning speed of 50 mm/s and relatively low laser power of 3000 W was adopted, optimal microstructures and properties were obtained. Too high a scanning speed of 60 mm/s could not realize metallurgical bonding between the HEA matrix and diamond. At high laser power and low scanning speed, the diamond graphitized and lost the crystal shape, and diffused C reacted with Cr and Mo to form carbides.
- With the increase of diamond content, the microstructures of the composite coating were obviously refined, and the wear resistance was improved. However, the high content (20 wt.%) aggravated the incompatibility of coating microstructures, resulting in microcracks. When the diamond content was 15 wt.%, optimal hardness and wear resistance were obtained.

Author Contributions: Conceptualization, W.Z.; methodology, W.Z.; validation, W.Z. and Y.P.; formal analysis, W.Z. and Y.P.; investigation, W.Z.; resources, Y.L.; data curation, M.Z., H.W. and S.L.; writing—original draft preparation, M.Z. and H.W.; writing—review and editing, W.Z. and Y.P.; supervision, Y.L.; project administration, W.Z. and Y.L.; funding acquisition, Y.L. All authors have read and agreed to the published version of the manuscript.

Funding: This research was funded by the project of State Key Laboratory of Powder Metallurgy (CSU 621011808), the National Natural Science Foundation of China (51731006), the Natural Science Foundation of Hunan Province (2020-2022 CSUZHANGWEI), the introduction projects of high-end foreign experts in 2019 for strategic science and technology development, Ministry of Science and Technology of the China (No: G20190010113), the Starting Research Fund of Nanjing Agricultural University (RCQD-1803).

Acknowledgments: The authors wish to acknowledge the financial support of State Key Laboratory of Powder Metallurgy (CSU 621011808), the National Natural Science Foundation of China (51731006), the Natural Science Foundation of Hunan Province (2020-2022 CSUZHANGWEI), the introduction projects of high-end foreign experts in 2019 for strategic science and technology development, Ministry of Science and Technology of the China (No: G20190010113), the Starting Research Fund of Nanjing Agricultural University (RCQD-1803).

Conflicts of Interest: The authors declare no conflict of interest.

References

1. Fang, X.; Deng, F.; Zheng, R. *Advanced Superabrasives and Related Products*; Zhejiang University Press: Hangzhou, China, 2011.
2. Wright, D.N.; Engels, J.A. The environmental and cost benefits of using diamond wire for quarrying and processing of natural stone. *Ind. Diamond Rev.* **2003**, *4*, 16–24.
3. Wang, F.; Zhang, J.; Wang, Z. Current situation and development of diamond wire saw cutting technology. *Diam. Abras. Eng.* **2013**, *33*, 36–42.
4. Zhang, Z.; Liu, B. *Development of Diamond Rope Saw*; West-China Exploration Engineering: Changchun, China, 1995.
5. Li, H.; Feng, Y.; Yang, L. Hybrid diamond rope saw. CN Patent CN203305381U, 27 Noverber 2013.
6. Weng, F.; Huang, G.; Huang, H. Study on sawing forces of sintered diamond wire-saw in granite cutting. *Diam. Abras. Eng.* **2007**, *01*, 25–27.
7. Huang, X. Design of Deep-Water Diamond Wire Saw Machine and Research on Life Prediction of Beaded Wire. Master's Thesis, Harbin Engineering University, Harbin, China, 2016.
8. Wang, C.; Liu, H.; Zhang, X. Effect of laser power and heat treatment process on microstructure and property of multi-pass Ni based alloy laser cladding coating. In Proceedings of the SPIE-The International Society for Optical Engineering, Beijing, China, 16 November 2010. [CrossRef]
9. Toyserkani, E.; Khajepour, A.; Corbin, S. *Laser Cladding*; CRC Press: Boca Raton, FL, USA, 2005.
10. Sun, H.H.; Guo, M.H.; Meng, F.L. Studies on hard faced overlay of diamond grits reinforced Ni-based alloy fabricated by laser cladding. *Trans. Indian Inst. Met.* **2016**, *69*, 1369–1376. [CrossRef]
11. Rommel, D.; Scherm, F.; Kuttner, C. Laser cladding of diamond tools: Interfacial reactions of diamond and molten metal. *Surf. Coat. Technol.* **2016**, *291*, 62–69. [CrossRef]
12. Iravani, M.; Khajepour, A.; Corbin, S. Pre-placed laser cladding of metal matrix diamond composite on mild steel. *Surf. Coat. Technol.* **2012**, *206*, 2089–2097. [CrossRef]
13. Yan, X.H.; Li, J.S.; Zhang, W.R. A brief review of high-entropy films. *Mater. Chem. Phys.* **2017**, *210*, 12–19. [CrossRef]
14. Vaidya, M.; Pradeep, K.G.; Murty, B.S. Bulk tracer diffusion in CoCrFeNi and CoCrFeMnNi high entropy alloys. *Acta Mater.* **2018**, *146*, 211–224. [CrossRef]
15. Verma, A.; Tarate, P.; Abhyankar, A.C. High temperature wear in CoCrFeNiCu$_x$ high entropy alloys: The role of Cu. *Scr. Mater.* **2019**, *161*, 28–31. [CrossRef]
16. Zhang, W.; Zhang, M.; Peng, Y. Effect of Ti/Ni coating of diamond particles on microstructure and properties of high-entropy alloy/diamond composites. *Entropy* **2019**, *21*, 164. [CrossRef]
17. Huang, F. An investigation on microstructure and properties of Ni-based alloy by laser cladding and laser cladding forming. Ph.D. Thesis, Jilin University, Changchun, China, 2011.
18. Zou, Q. Analysis of structures and surface states of nanodiamond and its treated methods. Master's Thesis, Yanshan University, Qinhuangdao, China, 2004.
19. Zhang, W.; Zhang, M.; Peng, Y. Interfacial structures and mechanical properties of a high entropy alloy-diamond composite. *Int. J. Refract. Met. Hard Mater* **2020**, *86*, 105109. [CrossRef]
20. Li, Z. Interstitial equiatomic CoCrFeMnNi high-entropy alloys: carbon content, microstructure, and compositional homogeneity effects on deformation behavior. *Acta Mater.* **2019**, *164*, 400–412. [CrossRef]

21. Liu, W.H.; Lu, Z.P.; He, J.Y.; Luan, J.H.; Wang, Z.J.; Liu, B.; Liu, Y.; Chen, M.W.; Liu, C.T. Ductile CoCrFeNiMo$_x$ high entropy alloys strengthened by hard intermetallic phases. *Acta Mater.* **2016**, *116*, 332–342. [CrossRef]
22. Peng, Y.; Zhang, W.; Mei, X. Microstructures and mechanical properties of FeCoCrNi-Mo High entropy alloys prepared by spark plasma sintering and vacuum hot-pressed sintering. *Mater. Today Commun.* **2020**, *24*, 101009. [CrossRef]
23. Zhang, M.; Peng, Y.; Zhang, W. Gradient distribution of microstructures and mechanical properties in a FeCoCrNiMo high-entropy alloy during spark plasma sintering. *Metals* **2019**, *9*, 351. [CrossRef]
24. Takeuchi, A.; Inoue, A. Classification of bulk metallic glasses by atomic size difference, heat of mixing and period of constituent elements and its application to characterization of the main alloying element. *Mater. Trans.* **2005**, *46*, 2817–2829. [CrossRef]

 © 2020 by the authors. Licensee MDPI, Basel, Switzerland. This article is an open access article distributed under the terms and conditions of the Creative Commons Attribution (CC BY) license (http://creativecommons.org/licenses/by/4.0/).

Article

Statistical/Numerical Model of the Powder-Gas Jet for Extreme High-Speed Laser Material Deposition

Thomas Schopphoven [1,*], Norbert Pirch [1], Stefan Mann [1], Reinhart Poprawe [1], Constantin Leon Häfner [1,2] and Johannes Henrich Schleifenbaum [1,3]

[1] Fraunhofer Institute for Laser Technology ILT, 52074 Aachen, Germany;
    norbert.pirch@ilt.fraunhofer.de (N.P.); stefan.mann@ilt.fraunhofer.de (S.M.);
    reinhart.poprawe@ilt.fraunhofer.de (R.P.); constantin.haefner@ilt.fraunhofer.de (C.L.H.);
    johannes.henrich.schleifenbaum@ilt.fraunhofer.de (J.H.S.)
[2] Chair for Laser Technology LLT, RWTH Aachen University, 52074 Aachen, Germany
[3] Chair for Digital Additive Production DAP, RWTH Aachen University, 52074 Aachen, Germany
*   Correspondence: thomas.schopphoven@ilt.fraunhofer.de; Tel.: +49-(0)241-8906-8107

Received: 18 March 2020; Accepted: 13 April 2020; Published: 22 April 2020

Abstract: Extreme high-speed laser material deposition, known by its German acronym EHLA, is a new variant of laser material deposition (LMD) with powdered additives. This variant's process control is unlike that of LMD, where the powder melts as it contacts the melt pool. In the EHLA process, the laser beam melts the powder above the surface of the substrate to deliver a liquid to the melt pool. At a given intensity distribution in a laser beam, the heating of powder particles in the beam path depends largely on the three-dimensional powder particle density distribution (PDD) and the relative position within the laser beam caustic. As a key element of a comprehensive numerical process model for EHLA, this paper presents a statistical/numerical model of the powder-gas jet, as previously published in Experimentelle und modelltheoretische Untersuchungen zum Extremen Hochgeschwindigkeits-Laserauftragschweißen. The powder-gas jet is characterized experimentally and described with a mathematical model. This serves to map the PDD of the powder-gas flow—and particularly the particle trajectories for different grain fractions—as well as the powder mass flows and carrier and inert gas settings, to a theoretical model. The result is a numerical description of the particle trajectories that takes into account the measured particle size distribution with calculations made on the assumption of a constant particle velocity and linear trajectories of the particles.

Keywords: extreme high-speed laser material deposition (EHLA); laser material deposition (LMD); coaxial powder nozzle; coating; additive manufacturing; numerical simulation

1. Introduction

Laser material deposition (LMD) is a process by which a laser beam creates a melt pool on the surface of a substrate while a powder nozzle delivers a powdered additive to the processing point–that is, into the melt pool. To produce a metallurgically bonded layer that is free of defects, the laser power, feed rate, powder mass flow and other process parameters have to be configured so as to apply sufficient process heat to trigger a suitable temperature–time cycle for the base material and additive. This application of heat goes to create a melt pool on the substrate surface and ensures that the additive melts fully. The relative motion of the laser beam and/or substrate leaves a dense, metallurgically bonded weld bead after the melt pool solidifies. A coating may be applied over a wider area by overlapping several weld beads. Multiple layers may be deposition-welded on top of one another, so the process may also be used for repair, additive manufacturing and hybrid additive manufacturing–that is, additive manufacturing on top of cast, forged or otherwise conventionally made base geometries, for example, to modify a part's geometry. Figure 1 diagrams the key factors influencing the process of forming a metallurgically bonded composite [1].

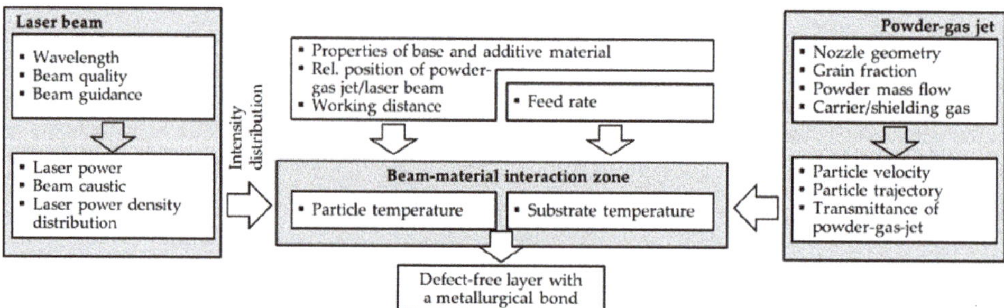

Figure 1. Diagram of the factors influencing powder-based laser material deposition (LMD) [1]. Reprinted with permission from Ref [1]. Copyright 2020, Fraunhofer Verlag.

In a heat balance equation for the melt pool's surface, the heat flow is the product of the thermalized, transmitted radiation less the heat flow induced by the temperature equalization between the powder particles and the melt pool's surface as well as the heat loss attributable to radiation. Equation (1) is a mathematical description of the heat balance at the melt pool's surface with the proportion of transmitted radiation I_{trans}, of the powder mass flow $\dot{m}_p$, the temperature-dependent specific heat capacity c_p of the material density ρ_p, the emissivity ε, the Stefan-Boltzmann constant σ, the temperature equalization between the temperature at the melt pool's surface T_S and the mean particle temperature T_p [2,3]:

$$\lambda\langle\nabla T, \vec{n}\rangle = \langle\vec{I}_{trans}[x(t), y(t)], \vec{n}\rangle - \dot{m}_p\rho_p c_p(T_S - T_P) - \varepsilon\sigma T^4 \tag{1}$$

Depending on the difference between the particle and melt pool temperatures $(T_s - T_p)$, the second term is a source or loss term for which a corresponding heat of fusion must be considered in the event that particle temperatures are lower than the melting temperature T_M, [2]. A distinction between two different process control strategies is to be made depending on the temperature of particles before they enter the melt pool: For one, there is conventional LMD with particle temperatures that are predominantly lower than the melting temperature $T_p < T_M$, or $(T_s - T_p) > 0$. For the other, there is the extreme high-speed laser material (EHLA) process in which a large share of the particles are to be fully melted by the laser beam before they arrive in the melt pool $T_p > T_M$, or $(T_s - T_p) < 0$ [1,4–7].

1.1. Process Ccontrol in Laser Material Deposition

Figure 2 illustrates the principle of the conventional LMD process with a continuous coaxial powder nozzle. The interaction time between powder particles and the laser beam is relatively short. The temperature of particles before they enter the melt pool is usually far lower than the melting temperature, so the particles are not melted fully until they arrive in the melt pool [8–11]. In the process control used to date, most of the laser beam's radiation is transmitted to the substrate, depending on the type of powder nozzles used, the grain fraction, the relative positions of the powder nozzle and the laser beam, and the working distance. This results in a relatively large heat-affected zone with dimensions ranging from a few tenths of a millimeter to one millimeter [1,11–17].

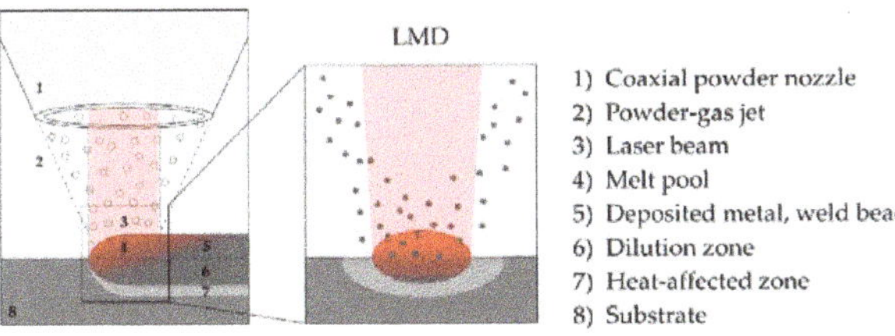

Figure 2. A diagram of the principle behind the LMD process [1]. Reprinted with permission from Ref [1]. Copyright 2020, Fraunhofer Verlag.

1.2. Process Control in Extreme High-Speed Laser Material Deposition

Unlike the process control for LMD, where the powder melts as it contacts the melt pool, in the EHLA process, the laser beam melts the powder above the surface of the substrate to deliver a liquid to the melt pool. Saving time otherwise required to melt the particles in the melt pool, this can increase the achievable feed rate from a few meters per minute to up to several hundred meters per minute [4–7]. Figure 3 illustrates the principle of the EHLA process. To this end, the ratio of transmitted laser radiation to the laser radiation absorbed by the powder particles has to be adjusted so that the heat input is sufficient to fully melt the particles already in the beam path and to create a melt pool on the surface of the substrate [7].

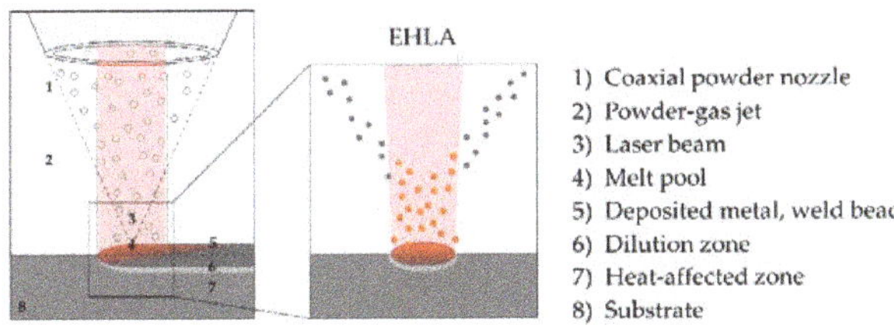

Figure 3. Diagram of the principle behind the extreme high-speed laser material deposition (EHLA) process [1]. Reprinted with permission from Ref [1]. Copyright 2020, Fraunhofer Verlag.

The key influencing factors on the heating of powder particles in the beam path above the melt pool are the powder particle trajectories, the particle velocity, the particle fraction and the relative position to the laser beam caustic. The additive's specific heat capacity, density, absorption coefficient and other thermophysical and optical properties also have an influence. The particles' trajectories and velocities determine the three-dimensional particle density distribution (PDD). However, as it stands given the current state of the art, there are no EHLA process control tools or methods available to describe the PDD for determining process-relevant influencing factors such as transmittance, particle heating in the beam path, substrate heating and track formation [1].

1.3. Modelling of the Powder-Gas Jet or Powder Particle Density Distribution

The formation of the powder-gas jet (particle velocity and trajectory) and thus the powder particle density distribution depends not only on the given nozzle's design and geometry. The powder grain fraction and the powder feed process parameters (powder mass flow, carrier and shielding gas volume flow) are also key determinants. The powder–gas flow is a dispersed, diluted two-phase solid flow consisting of the powder particles (disperse phase) and the carrier gas (continuous phase). Particle-to-particle and particle-to-nozzle wall collisions may occur depending on the degree of reciprocal action [18–22]. The particles are introduced into the upper part of the powder nozzle with the carrier gas so that their distribution is homogenized across the cross section by means of particle-to-wall collisions up to the nozzle outlet. The powder-gas flow exits the tip of the nozzle to enter an ambient medium with the shielding gas. Then turbulent mixing and swirling with the ambient medium causes the formation of a free jet [23–27]. Aligning the powder-gas jet on the substrate's surface transforms this original free jet into an impact jet. Contact with the substrate's surface reduces the gas flow velocity in the direction of the beam axis to zero [28–30]. The particles' velocities are reduced as a function of the particle following behavior, whereby the powder particles may still retain enough velocity to be fully immersed into the melt. The powder-gas flow is fully described by the volume-related conservation laws of mass, momentum and energy (the Navier-Stokes equations). However, the flow processes are so complex that at present they cannot be simulated without assuming simplified boundary conditions. Only the particles' trajectories and velocities prior to the nozzle's outlet are relevant to the interaction with the laser beam, which–depending on the given grain fraction and powder mass flow–determine the transmittance of the powder-gas jet and heating of the particles before they contact the substrate. Numerical and analytical models with varying depths of detail, scopes and assumptions for the mathematical definition of the powder-gas jet forming behind the nozzle tip are described in the literature. In most cases, the powder-gas jet is described numerically with the standard k-ε turbulence model for turbulent two-phase flows [31–35]. The k-ε model is a transport equation for a viscous vortex model in which the turbulent flow is described with the two parameters k for the turbulent kinetic energy and ε for the dissipation rate. The details of the two-phase flow within the nozzle are irrelevant to understanding the process and describing the particles' interaction with the laser beam–unless the goal is to learn what measures need to be taken to adapt the nozzle's design for a given purpose, for example, to achieve certain particle velocities or trajectories. Analytical descriptions of the powder-gas jet usually approximate the powder particle density distribution on the basis of a two-dimensional, radially symmetrical Gaussian distribution. To this end, the free parameters of the normal distribution pattern are defined according to the nozzle's geometry and a comparison to experimental results. Because the Gaussian distribution is radially symmetrical, these models are only valid forward of the point at which the powder-gas jet's hollow cone consolidates (the powder focus). The area prior to the point at which the annular powder distribution consolidates is not modeled at all [36–39] or approximated by partial jets [40–42]–either that, or it is assumed that the particle density distribution (PDD) of the powder-gas jet takes on a Gaussian shape immediately after the nozzle tip [36,43,44]. This paper presents a novel statistical/numerical model of the powder-gas jet as a key element of a comprehensive process model for the EHLA process. The underlying assumptions for this model have been derived from experimental observations and are based on images of the PDD captured at various levels with a laser-light sectioning device. It can serve to map the PDD of the powder-gas flow, the powder mass flows, and carrier and inert gas settings to a theoretical model.

2. Materials and Methods

2.1. Powdered Additive

The procedure for modeling PDD was demonstrated using powders sourced from the nickel-based superalloy material no. 2.4856 (similar to: Inconel 625, Special Metals Corporation, New Hartford, USA) with a nominal particle size distribution of −53 + 20 μm (trade name: MetcoClad 625F, made by Oerlikon Metco AG, Wohlen, Swiss) as an example. A scanning electron microscope served to investigate the powder particles' morphology. Figure 4 shows two images of the powder material taken with a scanning electron microscope (SEM) [1].

The powder particles are predominantly spherical. The satellites sporadically appearing on the surface are a result of the powder being produced by gas atomization. The particle size distribution was measured with a statistical image analysis at 20× magnification (Morphologi G3, Malvern Panalytical GmbH, Kassel, Germany). Figure 5 shows the measured relative F_{rel} and cumulative number-weighted frequency distributions F_{abs} against the equivalent diameter d_{CE}. The mean grain diameter is around 35 μm. In keeping with the number-weighted quantiles, around 96.5% of the measured particle diameters are within the manufacturer's specification [1].

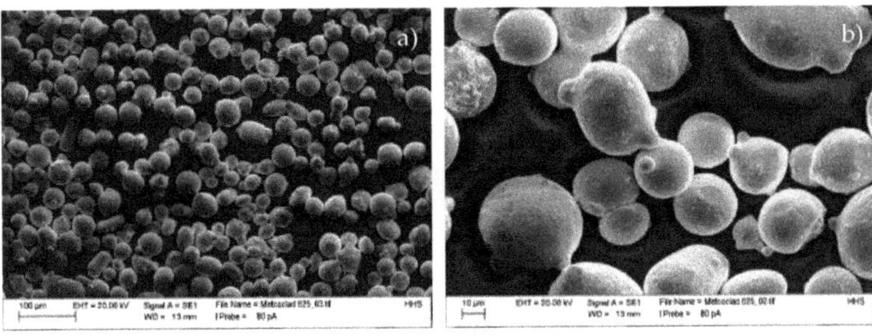

Figure 4. Scanning electron microscope (SEM) images of the powder with −53 + 20 µm grain fraction [1], (**a**) Overview; (**b**) Detail. Reprinted with permission from Ref [1]. Copyright 2020, Fraunhofer Verlag.

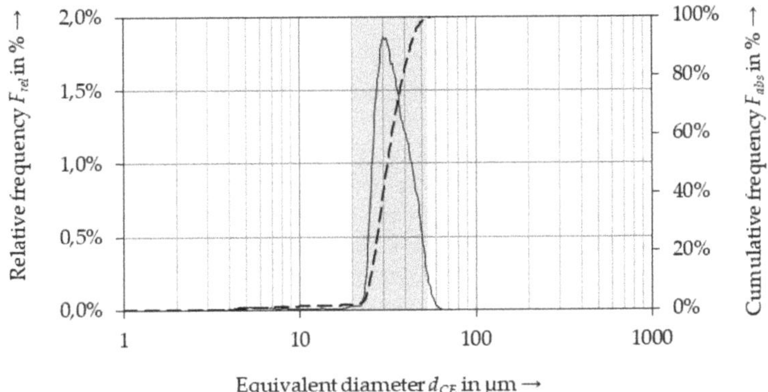

Figure 5. Relative F_{rel} and cumulative number-weighted frequency distributions F_{abs} against the equivalent diameter d_{CE}. Gray box: Manufacturer's specification with grain fraction −53 + 20 µm, mean grain diameter: 35.28 µm [1]. Reprinted with permission from Ref [1]. Copyright 2020, Fraunhofer Verlag.

2.2. Powder Feed

The powder was delivered by a type PF2/2 pneumatic powder feeder made by GTV Verschleißschutz GmbH (Luckenbach, Germany) with feed disc groove size of 0.6 × 5 mm D × W. A type GTV 108 electric mass flow controller made by Bronkhorst Mättig GmbH (Kamen, Germany) integrated into the powder feeder controlled the carrier gas's volume flow. A type GCR-C9SA-BA30 thermal mass flow controller made by Voigtlin Instruments AG (Aesch, Swiss) served to vary the inert gas volume flow. The carrier and shielding gas used for all experiments was argon with a primary pressure of four bar. The powder mass flow rate was measured gravimetrically for each powder grain fraction by weighing the powder quantity conveyed within one minute using a type LSM 200 precision scale (reproducibility 0.01 g) made by PCE GmbH (Meschede, Germany). Figure 6 shows the results of powder mass flow measurements $\dot{m}_p$ taken for one variation of the rotational speed U_P of the powder feed disc from 2 to 8 min^{-1} at a carrier gas volume flow of V_{FG} = 5 L/min.

The data points are the arithmetic mean of three measurements each. The investigation conducted for this paper used a constant carrier gas volume flow of 5 L/min and a shielding gas volume flow of 10 L/min. The required rotational speed of the powder feed disc UP was calculated with the given regression equations [1].

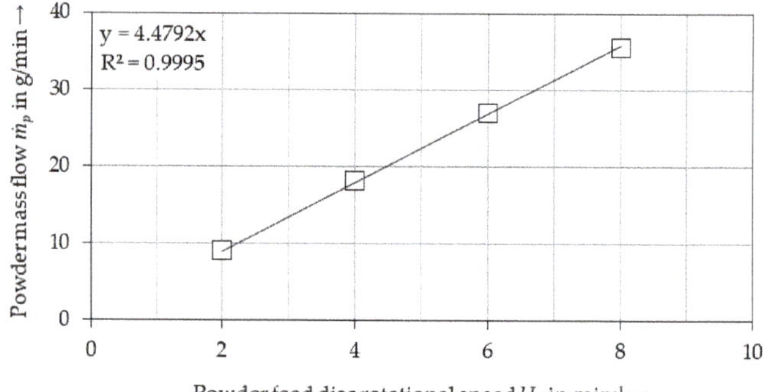

Figure 6. Measured powder mass flow $\dot{m}_p$ in relation to the powder feed disc rotational speed U_p with a grain fraction of −53 +20 μm [1]. Reprinted with permission from Ref [1]. Copyright 2020, Fraunhofer Verlag.

2.3. Powder Nozzle

An ILT-COAX 40-F powder nozzle (Fraunhofer ILT, Aachen, Germany) was used for these investigations. For this nozzle there is a splitter that divides the powder-gas jet into three partial jets that feed into an annular cavity sited between its inner and outer cones. The powder particles collide with the surface of the inner cone and flow with the carrier gas towards the nozzle's tip. This type of powder nozzle forms a hollow powder-gas cone. The powder-gas jet is focused on the tip of this cone (powder focus). Figure 7 diagrams this coaxial nozzle's structure. The width of the annular cavity's gap may be adjusted to the grain fraction and the powder mass flow by twisting the two interlocking cones of the coaxial powder nozzle. The larger the grain fraction and the greater the volume of the powder mass flow, the wider this gap has to be to prevent powder particles clogging the annular cavity between the inner and outer cones. Observations made during preliminary trials with the highest possible rotational speed of the powder feed disc of $U_P = 10$ min⁻¹ revealed that the annular cavity was no longer clogged at a gap width of 179 μm. The gap width of the annular cavity in the direction of flow was therefore around 179 μm for this investigation [1]. See Figure 7 for more on this.

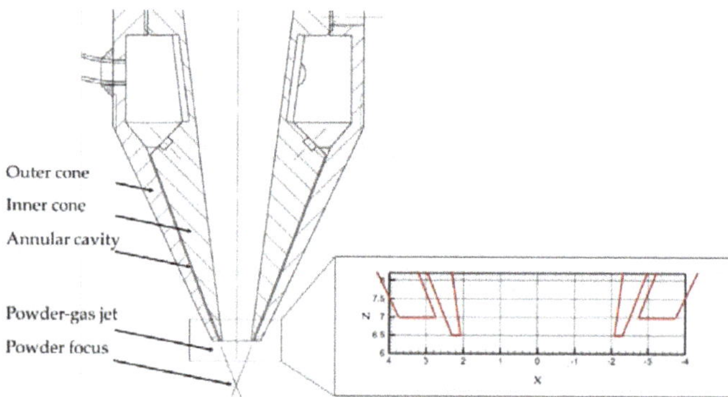

Figure 7. Schematic diagram of the powder nozzle and the annular cavity between the inner and outer cone of the continuous coaxial powder nozzle ILT-COAX 40-F [1]. Reprinted with permission from Ref [1]. Copyright 2020, Fraunhofer Verlag.

2.4. Powder-Gas Jet Ccharacterization

The metrological characterization of PDD was carried out with a laser light-sectioning method using an analytical instrument developed and patented by the Fraunhofer Institute for Laser Technology

ILT [45]. This method can measure the number and position of powder particles in the powder-gas flow level by level for various grain fractions, powder mass flows, and carrier and shielding gas settings. The powder-gas jet was illuminated laterally by a pulsed laser beam source developed at the Fraunhofer ILT (Aachen, Germany). Its wavelength is 810 nm. Optical components shaped the illuminating laser beam so that the height of the illuminating line in the direction of z was approximately 0.26 mm [1]. A type MC 1362 high-sensitivity, high speed CMOS camera made by Mikrotron GmbH (Unterschleißheim, Germany) was mounted coaxially to the powder nozzle. This camera's measuring range is 256 × 256 pixels at a resolution of 50 px/mm; the pixel size is approximately 19.53 μm. The pulsed illumination with a pulse duration of 0.008 ms enabled the powder particle position to be determined precisely while preventing light trails along the particle trajectory caused by motion blur. On the left of Figure 8 is a diagram of this setup; pictured on the right is the device used to measure the powder-gas jet [1].

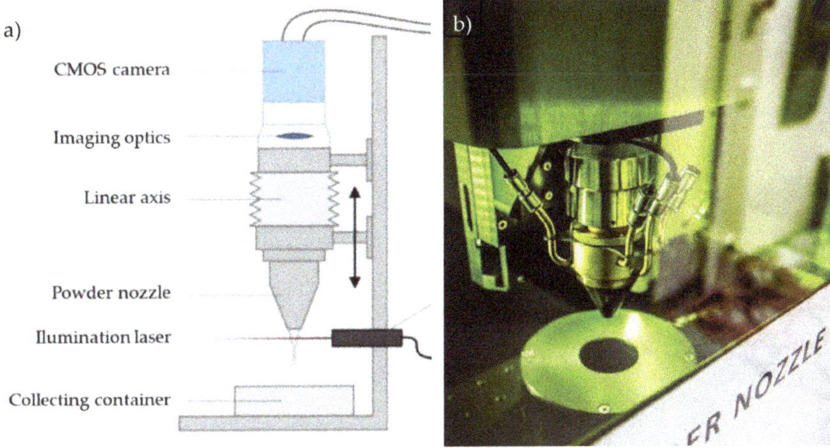

Figure 8. (**a**) Diagram of the experimental setup for measuring the powder-gas jet using the laser light-sectioning method. (**b**) Picture of the measuring device [1]. Reprinted with permission from Ref [1]. Copyright 2020, Fraunhofer Verlag.

Taking shots at a frame rate of 1000 frames per second, the camera can capture the number and position of powder particles at the respective level at a defined time. The powder nozzle was guided in alignment with the linear laser beam and photographed for three seconds at a time. This produced 3000 individual images per level with grayscale values ranging from 0 to 255 for each pixel of the camera chip. Figure 9 shows examples of individual images of the respective levels taken at a distance of 2 mm (E2) to 11 mm (E11) from the nozzle tip at a powder mass flow rate of 5 g/min [1].

The powder's annular distribution is visible when the 3000 individual images taken at each level are superimposed as shown in Figure 10. The powder distribution is no longer annular at a point below the powder focus level somewhere between E7 and E8. If the powder mass flow rate, powder grain fraction and illumination line height in the direction of z are known, then the powder particles' velocities can be approximated with a calculation based on the mean number of particles in the illuminated area [1].

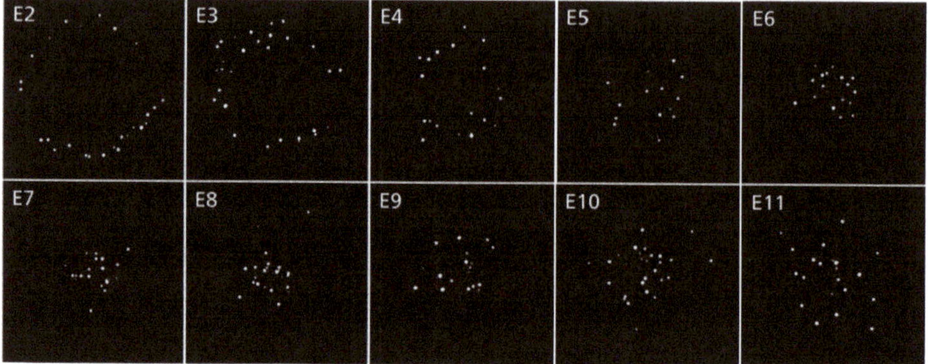

Figure 9. Selected examples of individual pictures of powder-gas jet measurements taken at a distance of 2 mm (**E2**) to 11 mm (**E11**). ILT-COAX 40-F powder nozzle. Parameters: grain fraction −53 + 20 μm, $\dot{m}_p$ = 5 g/min, V_{FG} = 5 L/min, V_{SG} = 10 L/min [1]. Reprinted with permission from Ref [1]. Copyright 2020, Fraunhofer Verlag.

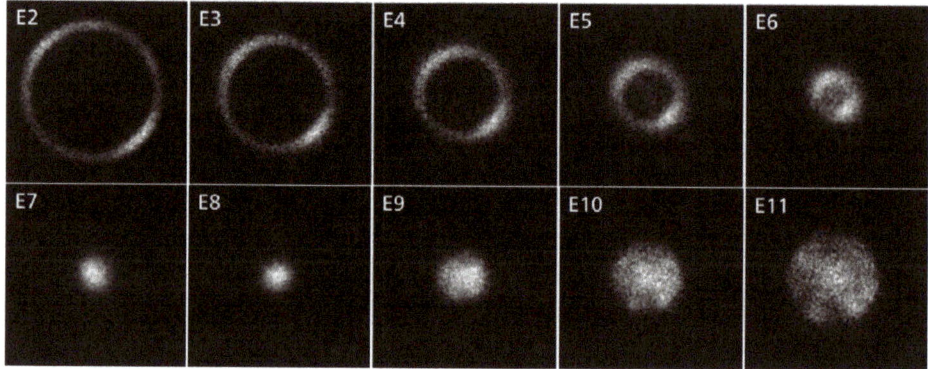

Figure 10. 3000 superimposed images, each of powder-gas jet measurements taken at a distance of 2 mm (E2) to 11 mm (E11). ILT-COAX 40-F powder nozzle. Parameters: grain fraction -53 + 20 μm, $\dot{m}_p$ = 5 g/min, V_{FG} = 5 L/min, V_{SG} = 10 L/min [1]. Reprinted with permission from Ref [1]. Copyright 2020, Fraunhofer Verlag.

3. Result

3.1. Powder Particle Count

A quantitative analysis can determine the number and position of particles in each image as depicted in Figure 9. Contiguous areas were designated as one particle each based on a threshold grayscale value. Figure 11 shows an example of particle identification on the basis of a threshold grayscale value for three individual images each at levels E3 and E4 at a powder mass flow of 5 g/min. The powder nozzle's conical design causes the expansion of the annular particle density at the nozzle outlet to decrease progressively until the powder focus is achieved. The particle density increases continuously as the distance from the powder nozzle tip grows. The greater the particle density, the higher the probability that the projections of two or more particles in the x-y plane overlap within the z measuring plane, and the fewer particles can be identified individually or the more particles are erroneously identified as one particle (particle–particle shading) [1].

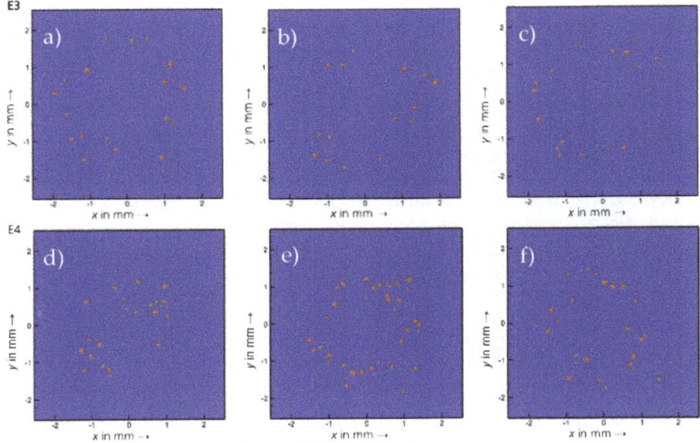

Figure 11. Diagrams of particle identification for each image based on a threshold for the grayscale value of the individual images. Parameters: grain fraction $-53 + 20$ µm, $\dot{m}_p = 5$ g/min, $V_{FG} = 5$ L/min, $V_{SG} = 10$ L/min, (**a–c**): Level E3; (**d–f**): Level E4 [1]. Reprinted with permission from Ref [1]. Copyright 2020, Fraunhofer Verlag.

Further factors influencing the particle density–that is, the measured number of particles–within a single image using this measuring method are [1]:

- The height of the illumination line or the measured level (here: 0.26 mm): The greater the height of the illumination line in the direction of z, the greater the number of simultaneously illuminated particles and, as a result, the higher the particle count within the measured level.
- Exposure time: The longer the exposure time (here: 0.008 ms), the greater the motion blur caused by light trails, which means that fewer particles can be identified.
- Resolution of the camera chip: The higher the resolution of the camera chip (here: 19.83 µm/pixel), the more likely it is that non-contiguous areas can be identified as individual particles.
- Distance of the measured level from the powder focus level: The shorter the distance between the measured level and the powder focus level, the greater the particle densities. The reason for this is that the particle count within the measured level decreases as the area of the powder-gas jet cross-section decreases.
- Grain fraction: The smaller the grain fraction, the higher the particle count at the same powder mass flow rate due to the proportionality to $1/r_p^3$, and the fewer the particles that can be identified individually.
- Particle velocity: The lower the particle velocity, the greater the number of simultaneously illuminated particles and the higher the particle count.
- Powder mass flow: The greater the powder mass flow, the higher the particle count in the measuring level at the same powder grain fraction.

Figure 12 shows an example of the overlap of several particles in the x-y plane within a single image. Starting at level E6, the number of particles detected for the investigated powder mass flows was significantly lower than at level E3. This is attributable to the increasing particle density along the beam axis caused by the decreasing area of the powder-gas jet cross-section and the resultant increasing overlap of particles [1].

For this reason, E3 serves in the following as a reference level to determine the mean particle count per level. The powder particle density distribution in the levels below the z plane is described with the help of a statistical model for the particle trajectories. See Section 3.3 for more on this. For example, the mean measured particle count per individual image is 68 for a powder mass flow rate of 15 g/min, a carrier gas volume flow rate of 5 L/min and an inert gas volume flow rate of 10 L/min [1].

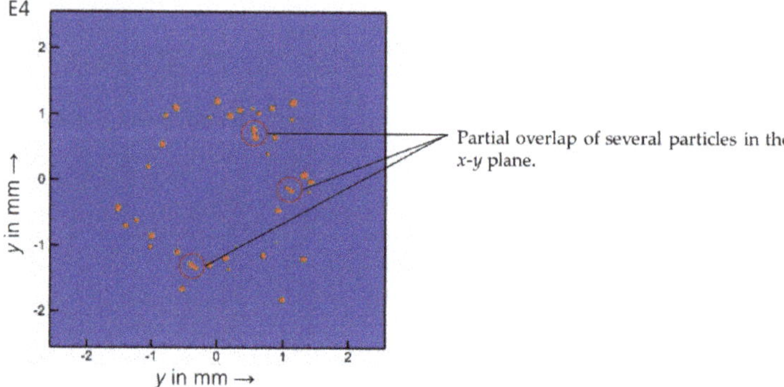

Figure 12. Diagram of the partial overlap in the x-y plane within a single image in level E4. Parameters: grain fraction −53 + 20 μm, $\dot{m}_p$ = 5 g/min, V_{FG} = 5 L/min, V_{SG} = 10 L/min [1]. Reprinted with permission from Ref [1]. Copyright 2020, Fraunhofer Verlag.

3.2. Approximation of the Particle Velocity

The number of particles n in a given volume can generally be expressed by the volume integral of the particle density distribution of the powder-gas jet $n(x,y,z)$, as expressed by Equation (2) [1]:

$$n = \int n(x,y,z)dxdydz. \tag{2}$$

Equation (4) expresses the number of particles n at the height of measuring level Δz [1]:

$$\frac{n}{\Delta z} = \int n(x,y,z)dxdy. \tag{3}$$

The powder mass flow $\dot{m}_p$ is equal to the mean particle number per level multiplied by the powder mass of a single particle m_p with the mean particle velocity $\bar{v}_p$ as expressed in Equation (5) [1]:

$$\dot{m}_p = \frac{n}{\Delta z} m_p \bar{v}_p. \tag{4}$$

This results in the mean particle velocity $\bar{v}_p$ as expressed by Equation (6) [1]:

$$\bar{v}_p = \frac{\Delta z\, \dot{m}_p}{n\, m_p} \tag{5}$$

Based on the aforementioned mean measured particle number per single image for a powder mass flow of 15 g/min, a carrier gas volume flow of 5 L/min and an inert gas volume flow of 10 L/min, the average particle velocity is $\bar{v}_p$ = 4.908 m/s [1].

3.3. Particle Propagation Model

Particle–particle shading yields excessive particle densities (see Section 3.1) that skew the particle count so their numbers cannot be measured with sufficient accuracy at every level. Pictures of the powder-gas jet taken with a high-speed camera show that the trajectories of the particles after they exit the nozzle aperture are rectilinear until they make contact with the substrate. This allows for an explicit representation of the particle trajectories and velocities in a statistical model. High-speed images of the powder-gas jet also confirm the assumptions that the particle velocity is constant and the propagation is linear. Figure 13 shows an example of a high-speed, multiple-exposure image of the powder-gas jet at a powder mass flow rate of 5 g/min, which is in line with the investigated process parameter range. The picture was taken with a Photron FASTCAM SA5 camera made by VKT Video Kommunikation GmbH (Pfullingen, Germany). The points are on a line (rectilinear propagation) and separated by an approximately constant distance (constant particle velocity) [1].

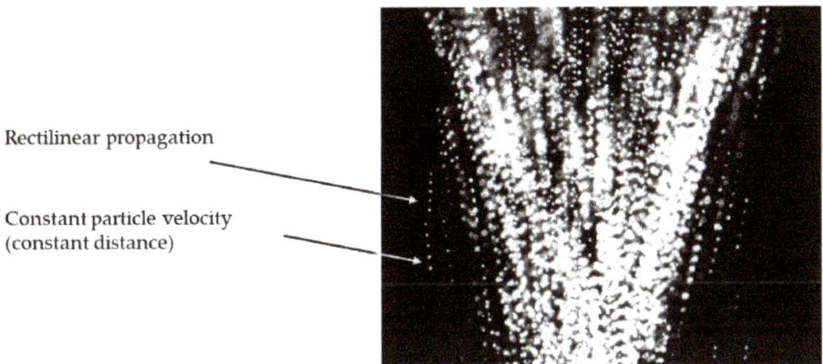

Figure 13. Images of the powder-gas jet taken at high speed with multiple exposures. Parameters: grain fraction $-53 + 20$ µm, $\dot{m}_p = 5$ g/min, $V_{FG} = 5$ L/min, $V_{SG} = 10$ L/min [1]. Reprinted with permission from Ref [1]. Copyright 2020, Fraunhofer Verlag.

A statistical description of the particles' trajectories requires figures for the distribution of starting points and direction of flight at the nozzle aperture. A particle's direction is determined by a directional vector given by the nozzle's angle of inclination. It is inclined in two perpendicular directions (see the diagram on the left of Figure 14). An initial distribution with an ensemble of $5{\cdot}10^5$ particles is predetermined at the nozzle outlet surface. Constrained by the nozzle walls ($r_1 < r < r_2$), this distribution has a Gaussian shape with respect to r and is uniform with respect to Φ. The inclination angles in this initial distribution are predetermined with respect to β $N(\beta, \Delta\beta^2)$ and γ normally distributed $N(0, \Delta\gamma^2)$. The starting points at the nozzle aperture are evenly distributed around the annular gap. The number of particles per second derives from the powder mass flow and the mean particle diameter. The resultant particle density in the reference level is calculated based on this input data. This does not correspond to the measured distribution, but should be covered by a suitable choice of γ and β. See the diagram on the left of Figure 15 [1].

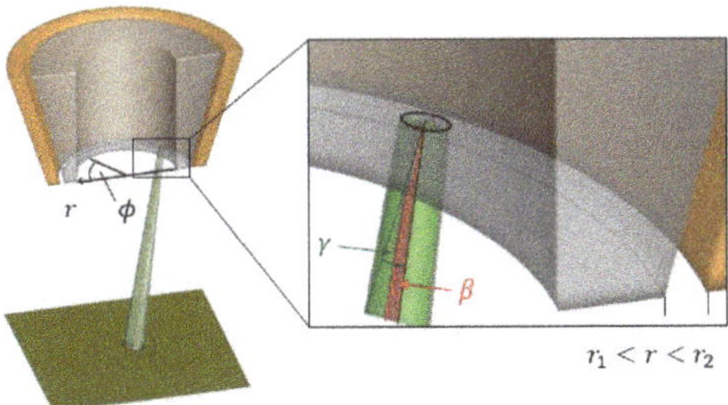

Figure 14. Diagram of the parameters for the particle propagation model. Left: Description of the powder particle density distribution with polar coordinates (r,Φ). Right: Model parameters for adjusting the angles inclination with respect to the z-axis: β and the x-y-axis: γ [1]. Reprinted with permission from Ref [1]. Copyright 2020, Fraunhofer Verlag.

In the next step, the calculated frequency distribution is normalized to one—that is, the maximum value is equal to one. A sequence of equally distributed random numbers in the range of [0,1] with the same length as the statistical particle ensemble is then generated. Each particle is assigned the corresponding random number from this sequence. If the random number for the particle lies above the normalized measured distribution, the particle is removed from the list. The remaining particles constitute a particle density at level E3 approximating the measured particle density. Figure 15 shows an example of the result of an adjustment made to model parameters β and γ and the filtering of trajectories for the particle density distribution of a powder-gas jet with a powder mass flow rate of 5 g/min at level E3 [1].

Non-uniform rational basis splines (NURBS) are also applied to smooth the measured particle density distribution. The graph on the left in Figure 16 shows an example of the modeled powder-gas jet with a powder mass flow of 5 g/min at a distance of up to 8 mm from the nozzle's tip. The center diagram shows the modeled particle density distributions. On the right is the composite of 3000 individual pictures of powder-gas jet measurements taken at levels E3 and E8 with the laser-light sectioning method. The result is a numerical description of the particle trajectories that takes into account the measured particle size distribution with calculations made on the assumption of a constant particle velocity and linear trajectories of the particles [1].

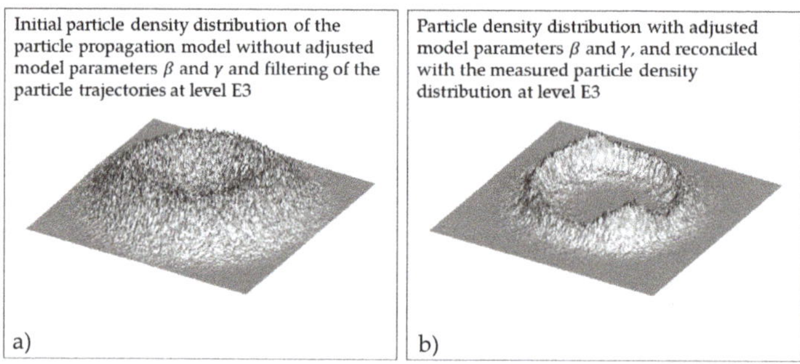

Figure 15. (**a**) Particle density distribution with a suitable selection of model parameters β and γ at level E3 and on the basis of the initial distribution, which is Gaussian with respect to r and uniform with respect to φ. (**b**) Particle density distribution after a statistical filtering of particle trajectories [1]. Reprinted with permission from Ref [1]. Copyright 2020, Fraunhofer Verlag.

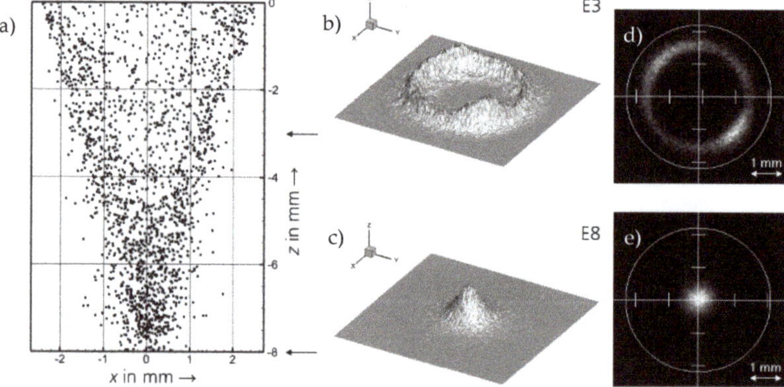

Figure 16. (**a**) Lateral view of the simulated powder-gas jet; (**b,c**): Simulated particle density distribution for levels E3 and E8. (**d,e**) 3000 superimposed images of measurements for levels E3 and E8. Parameters: grain fraction −53 + 20 μm, $\dot{m}_p$ = 5 g/min, V_{FG} = 5 L/min, V_{SG} = 10 L/min [1]. Reprinted with permission from Ref [1]. Copyright 2020, Fraunhofer Verlag.

3.4. Powder Particle Density Distribution

The statistical model for particle trajectories can also serve to determine the particle density for areas in which the particle density cannot be measured directly because of the partial particle overlap in the x-y-plane resulting from the decreasing area of the powder-gas jet cross section. A probability density was determined for each level to enable comparisons of the powder-gas jet's particle density distribution along the beam's propagation trajectory at various powder feed settings. The integral of the density distribution was normalized to a value of 1 via dx, dy to this end. The powder particle density distribution in absolute units of particles per cubic millimeter can be calculated by multiplying the probability density by the height of the illuminating laser beam dz, and normalizing it to the number of particles per level. Figure 17 shows an example of the modeled probability density for the reference setting as a function of the working distance between $z = -3$ mm and $z = -14$ mm by levels (left) and by $x = $ const. (right). The powder particle density progresses from an annular distribution at the nozzle aperture to a centered distribution at the focus of the powder jet in the direction of the beam. The powder-gas jet widens continuously along the levels below the powder focus relative to the beam axis [1].

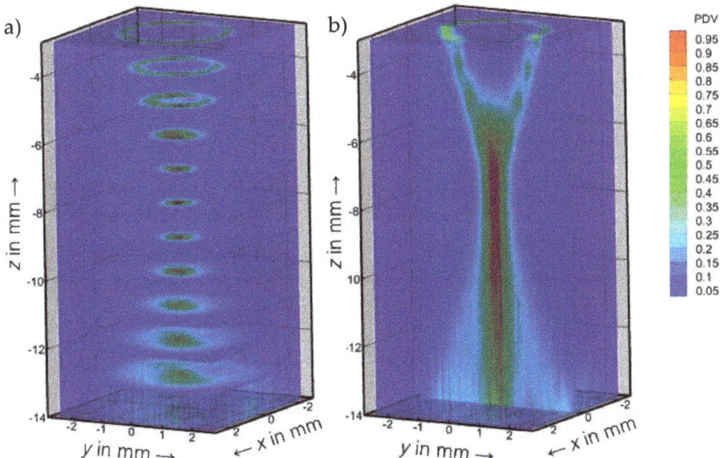

Figure 17. (**a**) level-wise probability density as a function of the working distance $7 = $ const., (**b**) level by level probability density as a function of the working distance for $\times = $ const. Parameters: grain fraction -53 + 20 μm, $\dot{m}_p = 15$ g/min, $V_{FG} = 5$ L/min [1]. Reprinted with permission from Ref [1]. Copyright 2020, Fraunhofer Verlag.

Pictured on the left of Figure 18 are lateral views of examples of the modeled powder-gas jet for different grain size distributions. On the right side are coaxial views of the powder-gas jet up to a distance of 8 mm from the nozzle tip (E8) with a laser beam diameter of $d_L = 1.13$ mm at the processing level (red). A drastic increase in particle density is clearly visible despite the comparable probability densities. This is the result of the particle count increasing as the mean particle diameter decreases [1].

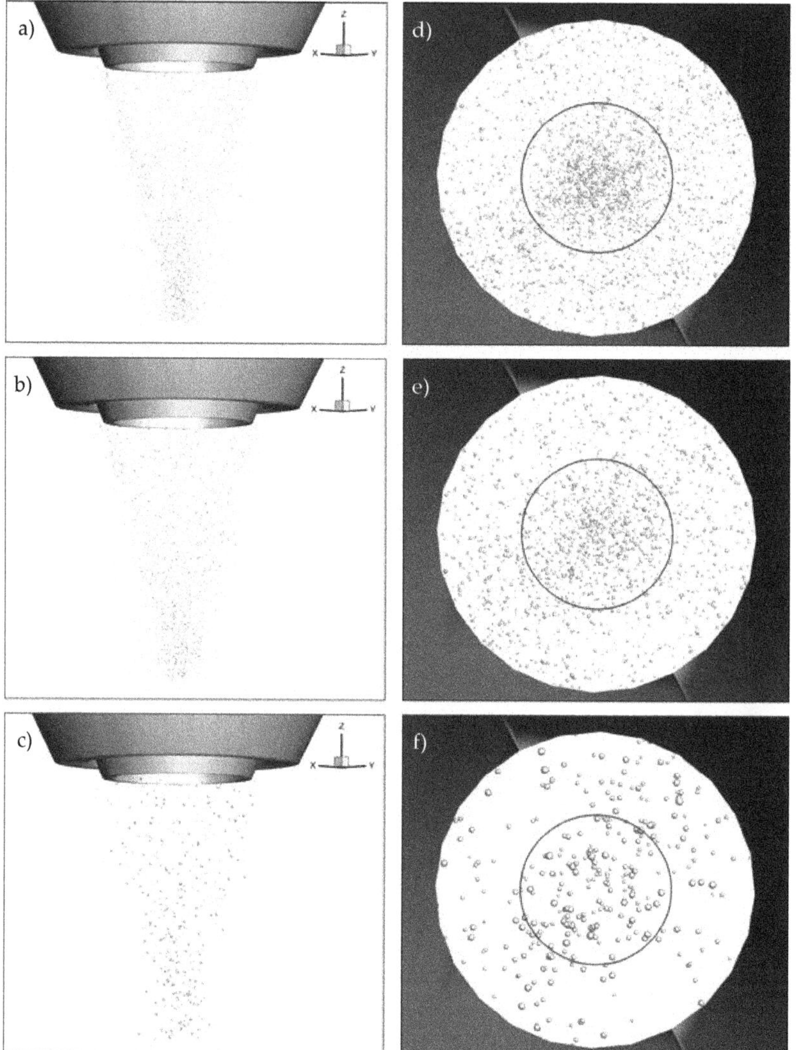

Figure 18. (**a–c**): Lateral views of the simulated powder-gas jet up to level E8. (**d–f**): Coaxial views of the powder-gas beam for level E8, red: laser beam. Grain fractions at the top: −45 + 11 μm, center: −53 + 20 μm, bottom: −90 + 45 μm. Parameters: $\dot{m}_p$ = 15 g/min, V_{FG} = 5 L/min, V_{SG} =10 L/min [1]. Reprinted with permission from Ref [1]. Copyright 2020, Fraunhofer Verlag.

4. Conclusions and Outlook

- The number and position of powder particles in the powder-gas flow can be determined quantitatively, level by level, with a measuring device based on the laser-light sectioning method.
- Pictures of the powder-gas jet taken with a high-speed camera serve to show that the trajectories of the particles after they exit the nozzle outlet are rectilinear and travel at a constant velocity until they contact the substrate. This allows for an explicit representation of the particle trajectories and velocities in a statistical model.

- The model theory-based description of the PDD presented in this paper is a key component of a comprehensive EHLA process model, as published in [1]. This serves to map the PDD of the powder-gas flow—and particularly the particle trajectories for different grain fractions—as well as the powder mass flows and carrier and inert gas settings to a theoretical model.

- In a further step the laser radiation will be characterized experimentally and mapped to the mathematical model.

- Based on this breakdown the interaction between the laser beam and the powder particles can be described. This enhances the understanding of the process-relevant influencing factors such as laser beam transmittance via the powder-gas jet, particle heating in the beam path, substrate heating and track formation.

- The transmitted laser intensity distribution and mean particle temperature are used as boundary conditions for the governing equations of the mathematical model for the track formation.

- This model will provide the means to assess the EHLA process's technological limits, for example, in terms of the achievable feed rate and track formation, and to push beyond these boundaries to effectively extend this process's range of potential manufacturing applications, particularly in surface engineering, but also in repair and additive manufacturing.

Author Contributions: Conceptualization, T.S.; methodology, T.S., N.P. and S.M.; software, N.P. and S.M.; T.S.; writing—original draft preparation, T.S.; writing—review and editing, N.P., S.M., R.P., C.L.H. and J.H.S.; visualization, N.P., and S.M.; supervision, R.P., C.L.H., and J.H.S.; project administration, T.S.; All authors have read and agreed to the published version of the manuscript.

Funding: This research received no external funding.

Acknowledgments: In adherence to the rules of good scientific practice it is stated, that the presented numerical model of the powder-gas jet for Extreme High-Speed Laser Material Deposition has been previously published in [1] in German language as part of a Ph.D. thesis. This publication serves to make the developed model available to a larger, international audience.

Conflicts of Interest: The authors declare no conflict of interest.

References

1. Schopphoven, T. Experimentelle und Modelltheoretische Untersuchungen zum Extremen Hochgeschwindigkeits-Laserauftragschweißen. Ph.D. Thesis, RWTH Aachen University, Aachen, Germany, 2020.

2. Pirch, N.; Linnenbrink, S.; Gasser, A.; Wissenbach, K.; Poprawe, R. Analysis of track formation during laser metal deposition. *J. Laser Appl.* **2017**, *29*, 022506. [CrossRef]

3. Pirch, N.; Niessen, M.; Linnenbrink, S.; Schopphoven, T.; Gasser, A.; Poprawe, R.; Schulz, W. Temperature field and residual stress distribution for laser metal deposition. *J. Laser Appl.* **2018**, *30*, 032503. [CrossRef]

4. Poprawe, R.; Hinke, C.; Meiners, W.; Eibl, F.; Zarei, O.; Voshage, M.; Willenborg, E.; Ziegler, S.; Schleifenbaum, J.H.; Gasser, A.; et al. Digital photonic production along the lines of industry 4.0. In *Laser Applications in Microelectronic and Optoelectronic Manufacturing (LAMOM) XXIII*; International Society for Optics and Photonics: San Francisco, CA, USA, February 2018; Volume 10519, p. 1051907.

5. Schopphoven, T.; Gasser, A.; Backes, G. EHLA: Extreme High-Speed Laser Material Deposition: Economical and effective protection against corrosion and wear. *Laser Technik J.* **2017**, *14*, 26–29. [CrossRef]

6. Schopphoven, T.; Gasser, A.; Wissenbach, K.; Poprawe, R. Investigations on ultra-high-speed laser material deposition as alternative for hard chrome plating and thermal spraying. *J. Laser Appl.* **2016**, *28*, 022501. [CrossRef]

7. Fraunhofer Gesellschaft zur Forderung der Angewandten Forschung. Extremes Hochgeschwindigkeitslaserauftragschweißverfahren. DE Patent DE 10 2011 100 456 B4, 4 May 2011.

8. Lin, J. Temperature analysis of the powder streams in coaxial laser cladding. *Opt. Laser Technol.* **1999**, *31*, 565–570. [CrossRef]

9. Liu, C.Y.; Lin, J. Thermal processes of a powder particle in coaxial laser cladding. *Opt. Laser Technol.* **2003**, *35*, 81–86. [CrossRef]

10. Ibarra-Medina, J.; Pinkerton, A.J. A CFD model of the laser, coaxial powder stream and substrate interaction in laser cladding. *Phys. Procedia* **2010**, *5*, 337–346. [CrossRef]

11. Ibarra-Medina, J.; Pinkerton, A.J. Numerical investigation of powder heating in coaxial laser metal deposition. *Surf. Eng.* **2011**, *27*, 754–761. [CrossRef]

12. Zhang, K.; Zhang, X.M.; Liu, W.J. Influences of Processing Parameters on Dilution Ratio of Laser Cladding Layer during Laser Metal Deposition Shaping. *Adv. Mater. Res.* **2012**, *549*, 785–789. [CrossRef]

13. Gasser, A.; Meiners, W.; Weisheit, A.; Willenborg, E.; Stollenwerk, J.; Wissenbach, K. Tailor-made surfaces and components—The use of laser radiation in surface technology and rapid manufacturing—Part 1. *Laser-Tech. J.* **2010**, *7*, 47–53. [CrossRef]

14. Beyer, E.; Wissenbach, K. *Oberflächenbehandlung mit Laserstrahlung*; Springer-Verlag: Berlin, Germany, 1998; Volume 7, pp. 243–251. [CrossRef]

15. Partes, K.; Seefeld, T.; Sepold, G.; Vollertsen, F. Increased efficiency in laser cladding by optimization of beam intensity and travel speed. In *Workshop on Laser Applications in Europe*; International Society for Optics and Photonics: Dresden, Germany, December 2005; Volume 6157, p. 615700.

16. Partes, K. High-Speed Coating with the Laser Beam. Ph.D. Thesis, Bremer Institut für Angewandte Strahltechnik, Bremen, Germany, 2008.

17. Peyre, P.; Aubry, P.; Fabbro, R.; Neveu, R.; Longuet, A. Analytical and numerical modelling of the direct metal deposition laser process. *J. Phys. D Appl. Phys.* **2008**, *41*, 025403. [CrossRef]

18. Liu, H.; He, X.; Yu, G.; Wang, Z.; Li, S.; Zheng, C.; Ning, W. Numerical simulation of powder transport behavior in laser cladding with coaxial powder feeding. *Sci. China Phys. Mech. Astron.* **2015**, *58*, 104701. [CrossRef]

19. Kussin, J. Experimental Studies on Particle Movement and Turbulence Modification in a Horizontal Channel with Different Wall Roughness. Ph.D. Thesis, Martin-Luther-Universität Halle-Wittenberg, Halle, Germany, 2004.

20. Kovalev, O.B.; Zaitsev, A.V.; Novichenko, D.; Smurov, I. Theoretical and experimental investigation of gas flows, powder transport and heating in coaxial laser direct metal deposition (DMD) process. *J. Therm. Spray Technol.* **2011**, *20*, 465–478. [CrossRef]

21. Pan, H.; Liou, F. Numerical simulation of metallic powder flow in a coaxial nozzle for the laser aided deposition process. *J. Mater. Process. Technol.* **2005**, *168*, 230–244. [CrossRef]

22. Pan, H.; Sparks, T.; Thakar, Y.D.; Liou, F. The investigation of gravity-driven metal powder flow in coaxial nozzle for laser-aided direct metal deposition process. *J. Manuf. Sci. Eng.* **2006**, *128*, 541–553. [CrossRef]

23. Elghobashi, S. On predicting particle-laden turbulent flows. *Appl. Sci. Res.* **1994**, *52*, 309–329. [CrossRef]

24. Wibel, W. Studies on Laminar, Transitional and Turbulent Flow in Rectangular Microchannels. Ph.D. Thesis, Technische Universität Dortmund, Dortmund, Germany, 2009.

25. Kravtsov, M.V. Resistance to the free steady-state motion of a sphere in a viscous medium. *J. Eng. Phys. Thermophys.* **1968**, *15*, 833–838.

26. Li, H.P.; Chen, X. Three-dimensional modeling of the turbulent plasma jet impinging upon a flat plate and with transverse particle and carrier-gas injection. *Plasma Chem. Plasma Process.* **2002**, *22*, 27–58. [CrossRef]

27. Haider, A.; Levenspiel, O. Drag coefficient and terminal velocity of spherical and nonspherical particles. *Powder Technol.* **1989**, *58*, 63–70. [CrossRef]

28. Gerold, J. Experimental and Numerical Investigation of Gas Free Jets. Ph.D. Thesis, Universität der Bundeswehr München, Neubiberg, Germany, 2015.

29. Richter, F.; Leder, A.; Dopheide, D.; Müller, H.; Strunck, V. (Eds.) Wechselwirkungen runder Düsenfreistrahlen mit ebenen Wänden. In Proceedings of the Fachtagung "Lasermethoden in der Strömungstechnik", Braunschweig, Germany, 7–9 September 2006.

30. Rajaratnam, N. *Turbulent Jets*; Elsevier: Amsterdam, The Netherlands, 1976; Volume 5.

31. Wen, S.Y.; Shin, Y.C.; Murthy, J.Y.; Sojka, P.E. Modeling of coaxial powder flow for the laser direct deposition process. *Int. J. Heat Mass Transf.* **2009**, *52*, 5867–5877. [CrossRef]

32. Lin, J. Numerical simulation of the focused powder streams in coaxial laser cladding. *J. Mater. Process. Technol.* **2000**, *105*, 17–23. [CrossRef]

33. Tabernero, I.; Lamikiz, A.; Ukar, E.; de Lacalle, L.L.; Angulo, C.; Urbikain, G. Numerical simulation and experimental validation of powder flux distribution in coaxial laser cladding. *J. Mater. Process. Technol.* **2010**, *210*, 2125–2134. [CrossRef]

34. Zhang, A.; Li, D.; Zhou, Z.; Zhu, G.; Lu, B. Numerical simulation of powder flow field on coaxial powder nozzle in laser metal direct manufacturing. *Int. J. Adv. Manuf. Technol.* **2010**, *49*, 853–859. [CrossRef]

35. Kovalev, O.B.; Kovaleva, I.O.; Smurov, I.Y. Numerical investigation of gas-disperse jet flows created by coaxial nozzles during the laser direct material deposition. *J. Mater. Process. Technol.* **2017**, *249*, 118–127. [CrossRef]

36. Yang, N. Concentration model based on movement model of powder flow in coaxial laser cladding. *Opt. Laser Technol.* **2009**, *41*, 94–98. [CrossRef]

37. Qi, H.; Mazumder, J.; Ki, H. Numerical simulation of heat transfer and fluid flow in coaxial laser cladding process for direct metal deposition. *J. Appl. Phys.* **2006**, *100*, 024903. [CrossRef]

38. Liu, S.; Zhang, Y.; Kovacevic, R. Numerical simulation and experimental study of powder flow distribution in high power direct diode laser cladding process. *Lasers Manuf. Mater. Process.* **2015**, *2*, 199–218. [CrossRef]

39. Diniz Neto, O.O.; Alcalde, A.M.; Vilar, R. Interaction of a focused laser beam and a coaxial powder jet in laser surface processing. *J. Laser Appl.* **2007**, *19*, 84–88. [CrossRef]

40. Pinkerton, A.J. An analytical model of beam attenuation and powder heating during coaxial laser direct metal deposition. *J. Phys. D Appl. Phys.* **2007**, *40*, 7323. [CrossRef]

41. Pinkerton, A.J.; Li, L. Modelling powder concentration distribution from a coaxial deposition nozzle for laser-based rapid tooling. *J. Manuf. Sci. Eng.* **2004**, *126*, 33–41. [CrossRef]

42. Pinkerton, A.J.; Li, L. A verified model of the behaviour of the axial powder stream concentration from a coaxial laser cladding nozzle. In *Proceedings of ICALEO*; Laser Institute of America: Orlando, FL, USA, 2002; Volume 2.

43. Lin, J.; Hwang, B.C. Coaxial laser cladding on an inclined substrate. *Opt. Laser Technol.* **1999**, *31*, 571–578. [CrossRef]

44. Liu, J.; Li, L.; Zhang, Y.; Xie, X. Attenuation of laser power of a focused Gaussian beam during interaction between a laser and powder in coaxial laser cladding. *J. Phys. D Appl. Phys.* **2005**, *38*, 1546. [CrossRef]

45. Fraunhofer Gesellschaft zur Forderung der Angewandten Forschung. Method and Device for Detecting a Particle Density Distribution in the Jet of a Nozzle. DE Patent DE102011009345B3, 25 January 2011.

 © 2020 by the authors. Licensee MDPI, Basel, Switzerland. This article is an open access article distributed under the terms and conditions of the Creative Commons Attribution (CC BY) license (http://creativecommons.org/licenses/by/4.0/).

Article

High Temperature Oxidation and Thermal Shock Properties of La$_2$Zr$_2$O$_7$ Thermal Barrier Coatings Deposited on Nickel-Based Superalloy by Laser-Cladding

Kaijin Huang [1,2,3,*], **Wei Li** [1], **Kai Pan** [2], **Xin Lin** [3] and **Aihua Wang** [1]

[1]   State Key Laboratory of Materials Processing and Die & Mould Technology, Huazhong University of Science and Technology, Wuhan 430074, China; liwei990628@hust.edu.cn (W.L.); ahwang@hust.edu.cn (A.W.)

[2]   Key Laboratory of Low Dimensional Materials & Application Technology, Xiangtan University, Xiangtan 411105, China; amrp@xtu.edu.cn

[3]   State Key Laboratory of Solidification Processing, Northwestern Polytechnical University, Xi'an 710072, China; xlin@nwpu.edu.cn

*   Correspondence: huangkaijin@hust.edu.cn; Tel.: +86-027-87543676

Received: 16 March 2020; Accepted: 7 April 2020; Published: 8 April 2020

Abstract: In order to reduce the difficulty and cost of manufacturing and improve the high temperature oxidation and thermal shock properties of nickel-based superalloy, a thin La$_2$Zr$_2$O$_7$ thermal barrier coating without bond coat was successfully prepared by laser-cladding using La$_2$Zr$_2$O$_7$ powders on a nickel-based superalloy substrate. Scanning electron microscopy (SEM) and X-ray diffraction (XRD) methods were used to characterize the microstructure of the coating. The high temperature oxidation and thermal shock properties of the coating were evaluated by the air isothermal oxidation method at 1100 °C for 110 h and thermal cycling method at 25~1100 °C, respectively. The results show that the coating is mainly composed of La$_2$Zr$_2$O$_7$ phase. The oxidation weight gain rate of the coating is about two-thirds of that of the substrate, and the first crack thermal shock lifetime of the coating is about 1.67 times of that of the substrate. The oxidation products of the coating are mainly Fe$_2$O$_3$, Cr$_2$O$_3$, NiCr$_2$O$_4$, Nb$_2$O$_5$ and La$_2$Zr$_2$O$_7$. The existence of La$_2$Zr$_2$O$_7$ phase in the coating is the main reason for the improvement of its oxidation resistance at 1100 °C and its thermal shock resistance at 25~1100 °C.

Keywords: laser-cladding; La$_2$Zr$_2$O$_7$ thermal barrier coating; Ni-based superalloy; high temperature oxidation; thermal shock

---

1. Introduction

With the development of modern industrial technology, nickel-based superalloys are widely used in aerospace, petrochemical and energy industries [1]. GH4169 is a commonly used precipitation strengthening nickel-based superalloy, which is resistant to high temperature oxidation, corrosion and radiation [2]. However, when its service temperature exceeds 1000 °C, the antioxidation properties of GH4169 at high temperature decline sharply, and a thermal barrier coating is usually required on its surface.

At present, the thermal barrier coating ceramic materials are Al$_2$O$_3$, SiO$_2$, yttria-stabilized zirconia (YSZ) and so on. Although the common YSZ ceramic materials can meet the thermophysical properties and phase stability requirements of the thermal barrier coating at around 1000 °C, YSZ is unable to meet the needs of the development of aeroengine, due to its easy phase transition above 1170 °C [3]. On the other hand, a new ceramic material for thermal barrier coating, La$_2$Zr$_2$O$_7$ (Lanthanum zirconate, abbreviations LZ), has attracted great attention due to its high melting point (2574 K), low thermal conductivity (1.6 W/(m·K)), low density (6.05 g /cm^3), low thermal expansion coefficient (9.1 × 10^{-6}

/K), no phase transition between room temperature and melting point, and no oxygen pentration [4]. For example, Doleker et al. [5] found that the high temperature oxidation at 1000°C and thermal shock at room temperature ~ 1150 °C of the LZ/YSZ double ceramic layer were better than that of the single layer YSZ coating when the bond coat was all CoNiCrAlY. Satpathy et al. [6] found that LZ and YSZ had good long-term physical compatibility and better thermal insulation after the preparation of nanometer LZ/YSZ double ceramic layer, by testing the durability of an aircraft engine. Wang et al. [7] found that thermal spraying of n-LZ /YSZ double ceramic coating improved the thermal insulation effect by about 35% compared with n-YSZ single coating. The thermal shock property at 1000 °C was twice as high as that of the n-YSZ coating. The oxidation resistance of two-phase ceramic layer was better than that of single-phase ceramic layer at any temperature. When oxidized at 1200 °C for 400 h, the weight of the dual-phase ceramic layer did not change significantly. Bobzin et al. [8] prepared YSZ, LZ and LZ/YSZ double ceramic layers by using electron beam—physical vapor deposition (EB-PVD). After oxidation at 1300 °C for 25 h, they found that the LZ was prone to fall off, while YSZ had no mixed oxides compared with LZ/YSZ.

It is well-known that the main practical methods for preparing thermal barrier coatings are plasma spraying (PS) and EB-PVD. The former has the advantages of a simple, economical and practical, good thermal insulation effect, but poor stress tolerance and corrosion resistance. The latter can obtain a coating with better aerodynamic and high temperature properties, but the equipment is expensive, complicated and inefficient. To solve the problems of poor oxidation resistance and short coating lifetime caused by high porosity and crack of plasma spraying coatings, the preparation of thermal barrier coatings by laser re-melting or laser-cladding has been studied. A large number of research results [9–19] in the world show that the laser re-melting of plasma spraying thermal barrier coatings can obtain dense columnar structures of epitaxial growth that are not available in plasma spraying coatings, thus improving the strain tolerance and thermal shock properties of the coatings. Laser-cladding can automatically stratify the composition and columnar structure of the gradient thermal barrier coatings, thus improving the high temperature oxidation and thermal shock properties of the coatings. Through the optimization of the laser process parameters and the reasonable design of the composition and properties of the coating system, the thermal barrier coating properties are better than that of plasma spraying and close to that of EB-PVD. In the process of preparing thermal barrier coating by laser-cladding, there are some reports about the use of bond coat [17–19] and not using bond coat [16,20–23]. Among them, the bond coat is mainly MCoCrAlY (M=Ni, Co, Ni–Co alloy), which mainly plays a bonding role and alleviates the difference of thermal expansion coefficient between the ceramic layer and substrate. However, when the thermal barrier coating temperature exceeds 1150 °C, the MCoCrAlY bond coat will oxidize rapidly. As the oxide film is gradually thickened, there will be cracks between the ceramic layer and the bond coat, or even falling off. The ceramic layer without bond coat mainly uses pure ceramic powder [20] or ceramic-based mixed powder [16,21–23]. Obviously, the thermal barrier coating with a bond coat must be prearranged with a bond coat, and then the ceramic layer can be prepared by laser-cladding. The preparation process is complex, and the advantage is that thick ceramic thermal barrier coating can be prepared. Of course, the thermal barrier coating without a bond coat can directly use laser-cladding to prepare the ceramic coating. The preparation process is simple, and the disadvantage is that the thin crack-free and no pores ceramic coating can only be prepared [20].

To our best knowledge, there are no literature reports on the preparation of LZ thermal barrier coating by laser-cladding, but there are a lot of literature reports on the preparation of LZ thermal barrier coating by PS and EB-PVD methods [24–26]. In order to use laser-cladding to prepare thin LZ thermal barrier coating (good thermal stability; long-term service temperature can exceed 1200 °C), this paper wants to try the laser-cladding without bond coat to prepare thin LZ coating, to reduce the preparation process and reduce the manufacturing cost.

In this paper, a thin $La_2Zr_2O_7$ thermal barrier coating without bond coat was prepared on the surface of GH4169 superalloy by laser preset cladding technology, and its high temperature oxidation and thermal shock properties at 1100 °C and 25~1100 °C were studied, respectively. The results show that the thin $La_2Zr_2O_7$ thermal barrier coating without bond coat has better high- temperature oxidation and thermal shock properties than GH4169 superalloy.

2. Materials and Methods

The laser cladding of $La_2Zr_2O_7$ (JCPDS 01-073-0444) powder (Figure 1) was carried out on a GH4169 nickel-based superalloy. The chemical composition of the used GH4169 superalloy is given in Table 1. The sample size used in the laser-cladding was 100 mm × 50 mm × 8 mm. The powder had a particle size of 37~75 μm and purity of 99.5%. The $La_2Zr_2O_7$ powder was preset on the surface of the GH4169 superalloy using 4% polyvinyl alcohol solution. The thickness of the preset layer was about 0.2 mm. The laser-cladding experiment was finished by using a LDF8000-60 type semiconductor laser-cladding system (Augsburg, Germany) (Figure 2). The preliminarily optimized laser-cladding parameters were as follows: laser power was 3500 W, laser scanning speed was 10 mm/s, laser spot size was 5 mm × 5 mm, and laser spot overlap rate was 30%. Figure 3 shows the macroscopic morphology of the sample after laser-cladding.

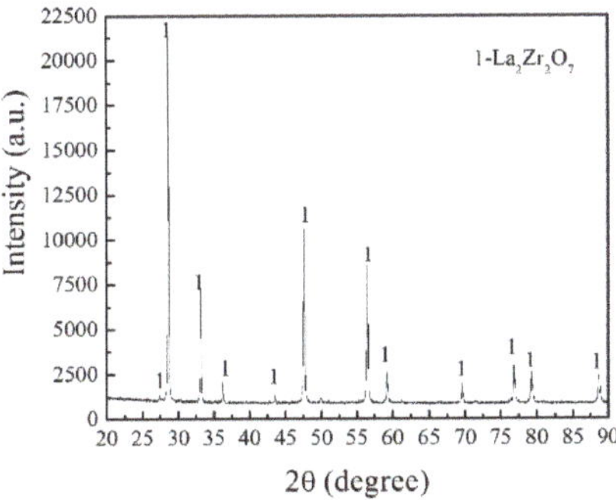

Figure 1. XRD patterns of $La_2Zr_2O_7$ powder.

Table 1. Chemical composition of the experimental GH4169 alloy.

Element	Cr	Ni	Nb	Al	Co	Ti	Fe
%	18.8	52.7	5.3	0.5	0.02	0.9	Bal.

Figure 2. Laser-cladding system.

Figure 3. Macroscopic morphology of the sample after laser-cladding.

The different phases of the coating were determined with the X-ray diffraction (XRD) technique using CuK_α radiation at 40 kV and 40 mA (X′ Pert PRO, Almelo, Netherlands, start position 2θ was 20.0066°, end position 2θ was 89.9856°, step size 2θ was 0.0130°, scan type was continuous, divergence slit size was 0.4354°). The microstructure of the cross-section coating etched with an aqua regia was examined by a Quanta 650 scanning electron microscope (SEM, FEI, Hillsboro, USA), with energy-dispersive X-ray spectroscopy (EDS, FEI, Hillsboro, USA).

The high temperature oxidation and thermal shock properties were tested according to Chinese aviation industry standard HB5258-2000 "Test method for the determination of oxidation resistance in steel and high temperature alloys" and Chinese aviation industry standard HB7269-96 "Quality inspection of thermal barrier coatings by thermal spraying", respectively. The test temperature of the high temperature oxidation was 1100 °C, and the test time was 2, 4, 8, 12, 16, 20, 30, 40, 50, 60, 70, 80, 90, 100 and 110 h, respectively. The detailed experimental process is as follows:

(1) A digital vernier caliper is used to measure the size of the coating and substrate samples and calculate the surface area S_0 of each sample.

(2) Washing, ultrasonic, drying alumina crucible.

(3) The dried crucible was placed in the M1230 type high temperature furnace and heated to 1150 °C for 30 min, then taken out and cooled to room temperature for weighing, and the crucible was burned to constant weight. Mass of the crucible was measured using an electronic analytical balance (Sartorius BS110, Goettingen, Germany) with an accuracy of 0.1 mg.

(4) After each sample is put into the crucible, its overall mass W_0 is weighed.

(5) The crucible containing the sample was placed in a high temperature furnace, which had been heated to 1100 °C for holding time.

(6) The crucible with the sample was taken out and cooled to room temperature and weighed. The crucible mass W_i with the sample was recorded.

(7) The mass difference ΔW_i ($\Delta W_i = W_i - W_0$) of each sample before and after oxidation together with the crucible was taken as the oxidation weight gain mass of the sample.

(8) Calculate the $\Delta W_i / S_0$ value.

(9) Repeat the above steps until all samples are completed.

(10) The $\Delta W_i / S_0$ - t curves of all samples were drawn as the isothermal oxidation kinetics curves of different samples at 1100 °C.

It should be pointed out that the sizes of the coating and the substrate samples were 8 mm × 8 mm × 1.3 mm and 8 mm × 8 mm × 8.7 mm, respectively. The heating rate for high temperature oxidation in the box-type resistance furnace was 10 °C/min. According to the requirements of HB5258-2000 standard, we have opened a hole of 12 mm in the door of the box-type resistance furnace, to make enough air enter the furnace chamber and maintain the oxidation atmosphere in the furnace chamber.

The thermal shock property was tested according to the following method: First, the sample (the coating or the substrate sample) was put into an electric furnace at 1100 °C and kept at this temperature for 15 min. Then, it was taken out and quickly cooled in the water at 25 °C, thus completing a thermal shock test. Repeat the process until a crack appears on the surface of the sample. At this point, the number of thermal shocks are denoted as the first crack thermal shock lifetime of the sample.

The surface morphologies of the oxidized specimens were observed using a scanning microscope (Quanta 650, FEI, Hillsboro, USA) with an energy spectrum.

3. Results and Discussion

3.1. Microstructure

Figure 4 shows the scanning electron microscopy morphology of the cross-section coating. When we adopted the measurement method in reference [27], the coating thickness was about 13~18 μm. It can be seen from Figure 4 that the outermost layer of the coating is relatively dense, but the density of the lower part of the coating needs to be improved (this should be achieved by further optimization of laser-cladding parameters), and the interface between the coating and the substrate is good except for a few cracks. This may be due to the large difference in the thermal expansion coefficient between $La_2Zr_2O_7$ and GH4169 (18.7×10^{-6}/K) and the rapid heating and cooling of the laser-cladding process. In Figure 4, energy-dispersive spectroscopy (EDS) 's results at area A show only La, Zr and O element, and the atomic percentages of the three elements are 15.4%, 18.3% and 66.3%, respectively. According to the XRD calibration results in Figure 5, it can be determined that the coating is indeed $La_2Zr_2O_7$ phase.

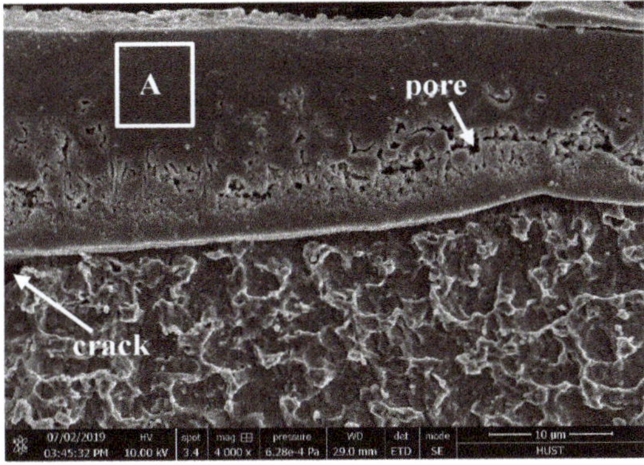

Figure 4. SEM image of the laser-clad $La_2Zr_2O_7$ coating.

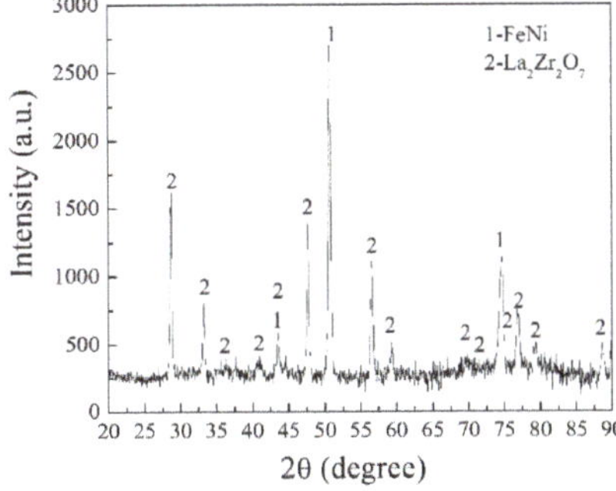

Figure 5. XRD patterns of the laser-clad coating.

Figure 5 presents the XRD patterns of the coating prepared by laser-cladding. The results show that the phases are $La_2Zr_2O_7$ (JCPDS 00-017-0450) and FeNi (JCPDS 00-018-0646) in the laser-clad coating. The $La_2Zr_2O_7$ phase should come from the laser melting and solidification of the $La_2Zr_2O_7$ powder, and the FeNi phase should come from the GH4169 substrate (Figure 6). For the $La_2Zr_2O_7$ phase, according to the diagram of La_2O_3 –ZrO_2 [28], the $La_2Zr_2O_7$ with pyrochlore structure exists in a rather large range temperature from melting point 2300 °C to room temperature. Therefore, when the $La_2Zr_2O_7$ powder is heated and melted by the laser, the $La_2Zr_2O_7$ phase does not decompose. When the laser is removed, the melted $La_2Zr_2O_7$ powder solidifies into the $La_2Zr_2O_7$ phase. For the FeNi phase, this may be related to the presence of a small number of pores in the lower part of the thin coating (Figure 4) that cause the X-ray to penetrate the coating to the substrate.

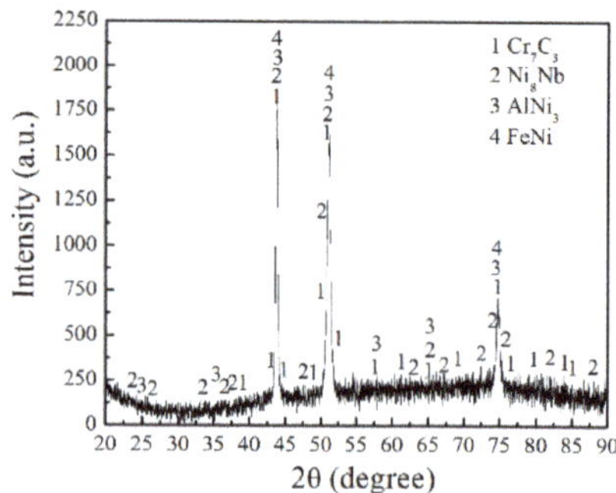

Figure 6. XRD patterns of the GH4169 substrate.

In addition, the ratios $I_{(222)}/I_{(440)}$ and $I_{(222)}/I_{(622)}$ of the peak height intensity of crystal planes (222) (at $2\theta = 28.59°$), (440) (at $2\theta = 47.56°$) and (622) (at $2\theta = 56.44°$) in Figure 1 are 2.03 and 2.68, respectively, while the ratios $I_{(222)}/I_{(440)}$ and $I_{(222)}/I_{(622)}$ of crystal plane (222) (at $2\theta = 28.66°$), (440) (at $2\theta = 47.69°$) and (622) (at $2\theta = 56.52°$) in Figure 5 are 1.10 and 1.48, respectively. This may be related to the preferential growth of the crystal planes (440) and (622) in $La_2Zr_2O_7$ phase, during the solidification of the laser-cladding. This phenomenon of preferential facet growth has been widely reported in the literature [29]. Different exposed crystal faces have different oxygen adsorption energies, resulting in different interactions between oxygen and LZ [30], which further affects the antioxidation property of the LZ coating at high temperature.

Moreover, compared with Figure 1, the peak positions of crystal planes (222), (440) and (622) in Figure 5 are shifted to the right. That is, these peak positions are correspondingly moved from 28.59°, 47.56° and 56.44° in Figure 1, to 28.66°, 47.69° and 56.52° in Figure 5, respectively. According to the Bragg diffraction formula $2d\sin\theta = n\lambda$, the increase of incident angle θ leads to the decrease of the crystal plane spacing d value, which further leads to the decrease of the lattice constant. In fact, the lattice constant of $La_2Zr_2O_7$ phase of the cubic crystal structure in Figure 1 is a = b = c = 1.0808 nm, and in Figure 5 is a = b = c = 1.07930 nm. This may be related to the doping of different alloying elements, such as Ni, Cr, Fe, Nb, etc. from the GH4169 substrate into the $La_2Zr_2O_7$ phase during the laser-cladding process. Since the ionic radii of Ni^{2+} (0.072 nm), Cr^{3+} (0.069 nm), Fe^{3+} (0.064 nm) and Nb^{5+} (0.070 nm) are all smaller than that of La^{3+} (0.106 nm) in $La_2Zr_2O_7$ phase; the lattice constant of the $La_2Zr_2O_7$ phase can be reduced when these alloying ions partially replace La^{3+} to form a substitution solid solution. Literature [28] also confirmed the existence of this phenomenon.

As can be seen from Figure 6, the GH4169 substrate is mainly composed of FeNi (JCPDS 00-003-1209), $AlNi_3$ (JCPDS 00-009-0097), Cr_7C_3 (JCPDS 00-036-1482) and Ni_8Nb (JCPDS 00-023-1274) phases. This is consistent with the results reported in the literature [31].

3.2. High Temperature Oxidation Properties

Figure 7 shows the oxidation weight gain rate of different samples at 1100 °C for different times. It can be seen from Figure 7 that the oxidation weight gain rate of the $La_2Zr_2O_7$ coating is smaller than that of the GH4169 substrate. This indicates that the $La_2Zr_2O_7$ coating can protect the GH4169 substrate from oxidation. The reason is that both the oxygen vacancy mobility and the oxygen ion conductivity of $La_2Zr_2O_7$ are very low [32], and $La_2Zr_2O_7$ is a ceramic material with extremely low oxygen permeability, which is basically no oxygen permeability material [4,33]. In addition, the oxidation weight gain rate of the $La_2Zr_2O_7$ coating is about two-thirds of that of the GH4169 substrate at 1100 °C for 110 h.

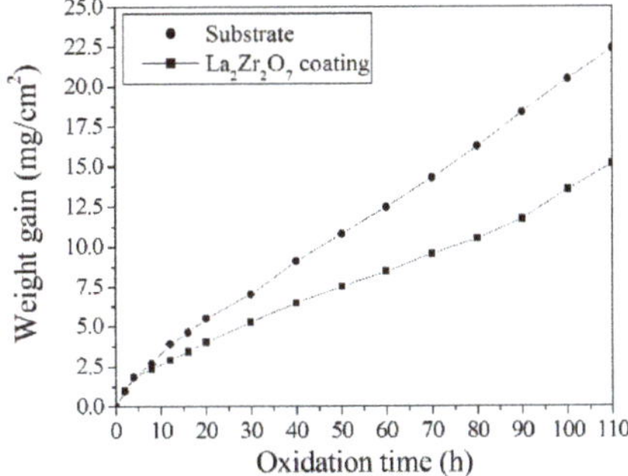

Figure 7. Oxidation weight gain rate of different samples at 1100 °C for different times.

Figures 8 and 9 show the XRD patterns of different samples after high temperature oxidation at 1100 °C for 110 h, respectively. As can be seen from Figure 8, after the oxidation at 1100 °C for 110 h, NiO (JCPDS 01-078-0643), Cr_2O_3(JCPDS 01-082-1484), Nb_2O_5(JCPDS 00-015-0166), $NiCr_2O_4$(JCPDS 01-075-0198), Fe_2O_3(JCPDS 01-076-1821) oxides were formed on the surface of the GH4169 substrate. This is consistent with the results reported in the literature [27]. During the oxidation process of the GH4169 substrate, dense Cr_2O_3 film, dense Al_2O_3 film and loose Nb_2O_5 film were firstly formed on the surface of the substrate. Although the dense Cr_2O_3 and Al_2O_3 oxide film can prevent the diffusion of oxygen element, the oxygen element can continue to diffuse into the interior, due to the loose Nb_2O_5 oxide film. Therefore, when the oxidation continues, NiO, Fe_2O_3 and $NiCr_2O_4$ oxides are formed [34]. It can be seen from Figure 9 that after the oxidation at 1100 °C for 110 h, in addition to maintaining part $La_2Zr_2O_7$ (JCPDS 01-073-0444) phase, oxides such as Cr_2O_3(JCPDS 00-038-1479), Nb_2O_5(JCPDS 00-015-0166), $NiCr_2O_4$(JCPDS 00-004-0763) and Fe_2O_3(JCPDS 01-076-1821) were formed on the coating surface. Among them, the last four oxides should come from the oxidation of the GH4169 substrate. There are three possible reasons for the five oxides: First, the thermal stability of $La_2Zr_2O_7$ phase is high, and it will not decompose at 1100 °C, so the $La_2Zr_2O_7$ phase will persist in the coating. Second, there are a small number of pores and cracks in the coating (Figure 4). Third, the $La_2Zr_2O_7$ coating demonstrates cracking or layered peeling, due to the difference of thermal expansion coefficient between the $La_2Zr_2O_7$ ceramic layer and the GH4169 substrate. During the long oxidation process at high temperature, oxygen will diffuse to the GH4169 substrate through the pores and cracks, and then cause partial oxidation of the GH4169 substrate and form the Cr_2O_3, Nb_2O_5, $NiCr_2O_4$ and Fe_2O_3.

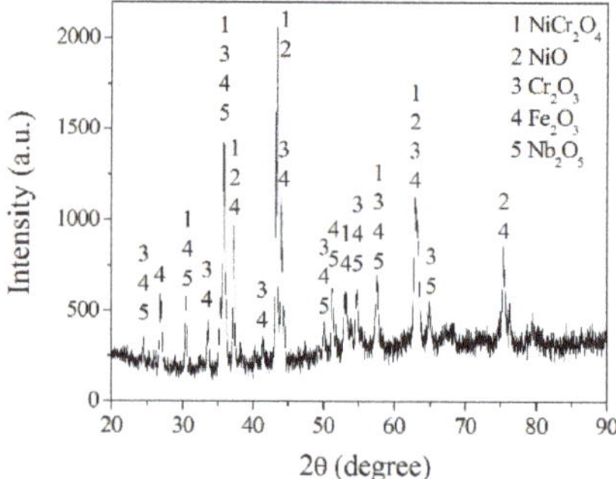

Figure 8. XRD patterns of the GH4169 substrate after high temperature oxidation at 1100 °C for 110 h.

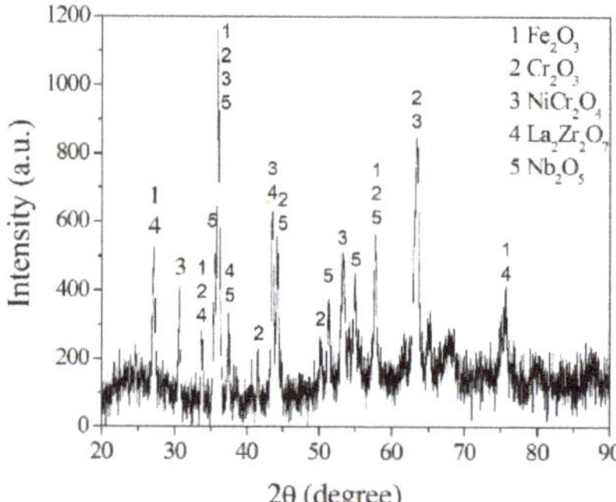

Figure 9. XRD patterns of the coating after high temperature oxidation at 1100 °C for 110 h.

Figure 10 as with Figure 11, Figure 10 shows the surface and cross-section morphologies of different samples after high temperature oxidation at 1100 °C for 110 h, respectively. Tables 2 and 3 are the EDS's.

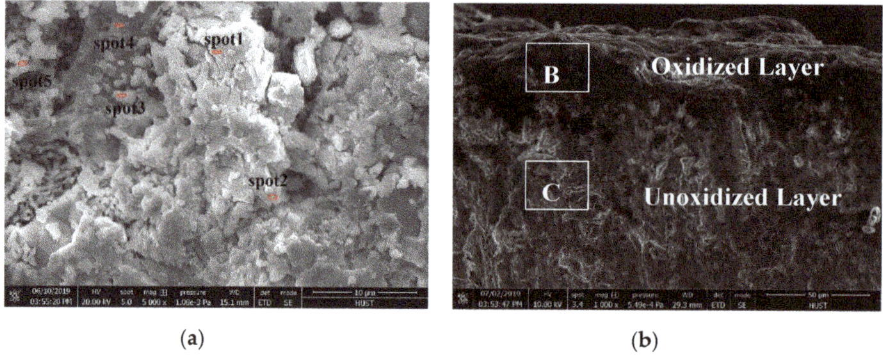

(a)　　　　　　　　(b)

Figure 10. Surface and cross-section morphologies of the oxidized substrate at 1100°C for 110 h, (**a**) Surface morphology; (**b**) Cross-section morphology.

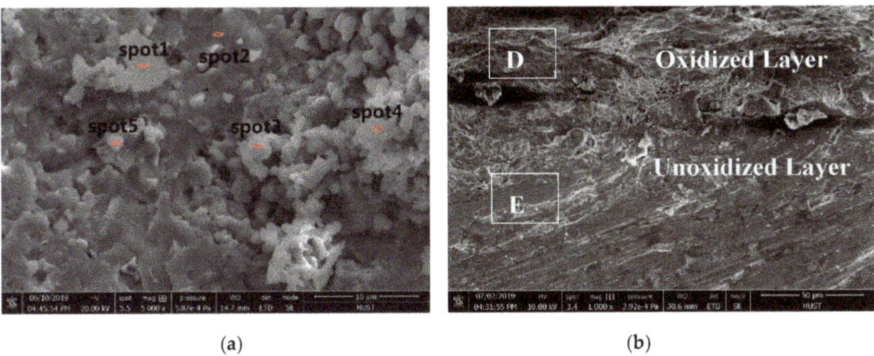

Figure 11. Surface and cross-section morphologies of the oxidized coating at 1100 °C for 110 h; (**a**) Surface morphology; (**b**) Cross-section morphology.

Table 2. EDS's results of GH4169 at different positions in Figure 10a (%).

Element	O	Ni	Nb	Cr	Fe	Ti	Co	Al
Spot1	38.7	32.8	16.0	7.4	5.1	0	0	0
Spot2	55.1	31.8	9.8	1.9	1.4	0	0	0
Spot3	52.1	32.9	8.5	4.4	2.1	0	0	0
Spot4	30.9	21.7	8.4	30.7	7.4	0.9	0	0
Spot5	53.7	34.2	7.9	2.9	1.3	0	0	0
Area B	62.1	18.2	6.9	1.8	6.3	0.8	0.3	3.6
Area C	0	22.9	25.0	33.9	12.4	2.6	0.1	3.1

Table 3. EDS's results of the coating at different positions in Figure 11a (%).

Element	O	Zr	La	Ni	Nb	Cr	Ti	Fe	Al	Co
Spot1	65.9	2.1	0.3	1.6	11.9	15.9	1.1	1.3	0	0
Spot2	57.2	1.8	1.0	5.8	18.7	12.1	0	3.0	0.4	0
Spot3	63.7	1.7	0.1	3.5	1.6	22.6	0.3	6.5	0	0
Spot4	66.8	1.6	0.2	1.7	9.6	12.5	2.7	3.4	1.5	0
Spot5	72.0	1.5	0.1	4.5	10.0	7.9	1.7	2.3	0	0
Area D	49.1	10.8	0.4	15.4	4.2	3.6	1.0	6.7	8.6	0.2
Area E	0	0	0	56.9	7.2	15.8	0.5	17.7	1.8	0.1

Combined with the XRD (Figures 8 and 9) and EDS's results (Tables 2 and 3) of the different oxidized samples, it can be seen that the oxidation products of the GH4169 substrate after oxidation mainly include NiO, Cr_2O_3, Nb_2O_5, $NiCr_2O_4$ and Fe_2O_3 phases, and the thickness of the oxide layer is about 40~45 µm (Figure 10b). The oxidation products of the coating after oxidation mainly include $La_2Zr_2O_7$, Cr_2O_3, Nb_2O_5, $NiCr_2O_4$ and Fe_2O_3 phases, and the thickness of the oxide layer is about 18 ~ 30 µm (Figure 11b).

In addition, it can be seen from Figure 7 and Table 3 that the oxidation mechanism of the $La_2Zr_2O_7$ coating and the substrate is basically the same. The difference is that the $La_2Zr_2O_7$ coating may first generate stress in the coating, due to the difference of thermal expansion coefficient between the $La_2Zr_2O_7$ ceramic layer and the GH4169 substrate in the thermal cycling process, which is called thermal mismatch stress, and then leads to the coating cracking or layered peeling. Once the $La_2Zr_2O_7$ coating appears cracking or layered peeling, the external oxygen will spread to the surface of the GH4169 substrate and undergo the same oxidation reaction as the GH4169 substrate. This indicates that the thin $La_2Zr_2O_7$ coating without bond coat prepared by laser-cladding can delay the oxidation reaction time of the GH4169 substrate to a certain extent, thus improving the oxidation resistance of the GH4169 substrate at high temperature. Of course, if a bond coat is present, the oxygen diffused from the outside will first react with the bond coat alloy, by oxidizing to form the thermal growth oxide (TGO) layer. Once the thickness of TGO layer increases to a certain amount, it will also crack due to stress mismatch, resulting in layer cracking or peeling [35]. Subsequently, the GH4169 substrate will also undergo oxidation reaction. From the high temperature oxidation process without bond coat and with bond coat, it can be seen that the coating with bond coat delays the oxidation of the GH4169 substrate longer than the coating without bond coat. However, the preparation process of the bond coat is an independent process, such as the plasma spraying process, which undoubtedly increases the manufacturing difficulty and cost. If the requirement of the coating lifetime is not very long, such as the thermal insulation layer for launching satellites and missile warheads, the manufacturing method of the coating without bond coat can be considered. This is also the significance of this study.

Moreover, Figure 7 is actually the oxidation kinetics curve of the GH4169 substrate and laser-clad $La_2Zr_2O_7$ coating, oxidizing 0–110 h at 1100 °C. According to Wagner theory [36], when the metal is oxidized at high temperature, it is possible to protect the metal only when a complete dense oxide film with good adhesion to the metal substrate is formed [37]. According to the theory, under ideal conditions, the relationship between the thickening of the oxide film and the oxidation time is parabolic:

$$\Delta W^2 = Kt \tag{1}$$

Oxidation weight gain is usually used to replace the increase in oxide film thickness. In formula (1), ΔW stands for oxidation weight gain, K stands for oxidation rate constant and t stands for oxidation time. However, in practice, the premise of formula (1) is usually unsatisfied, and the following modified formula is usually required:

$$\Delta W^n = Kt \tag{2}$$

In order to facilitate fitting, deformation is adopted in the calculation formula [38]:

$$lnt = nln\Delta W - lnK \tag{3}$$

The oxidation kinetics curves of the GH4169 substrate and the coating were fitted according to the formula (3) based on the data of Figure 7, and the n and K values were obtained, respectively (Table 4). As can be seen from Table 4, the n and K values of the coating are 1.53 and 0.47, respectively. The n and K values of the GH4169 substrate are 1.31 and 0.48, respectively. According to the judgment that "the larger the oxidation index is, the smaller the oxidation rate constant is, the denser the oxide film is, and the better the oxidation resistance at high-temperature is", the high temperature oxidation resistance of the coating is better than that of the GH4169 substrate. This is consistent with the experimental results (Figure 7).

Table 4. Oxidation rate constant K and oxidation index n of the coating and substrate.

Samples	Oxidation Index n	Oxidation Rate Constant K
$La_2Zr_2O_7$ coating	1.53	0.47
GH4169 substrate	1.31	0.48

3.3. Thermal Shock Properties

According to Chinese aviation industry standard HB7269-96, we obtained that the first crack lifetime of the coating and the substrate samples were 100 and 60, respectively. This indicates that the thermal shock property of the coating is better than that of the substrate, and the first crack lifetime of the coating is 1.67 times that of the substrate. This may be related to the low thermal expansion coefficient of $La_2Zr_2O_7$ phase in the coating and the high thermal expansion coefficient of the substrate. According to the thermal stress formula, $\sigma = \alpha \times \Delta T$, the thermal stress σ is small under the same temperature difference ΔT because the coefficient α of thermal expansion of $La_2Zr_2O_7$ phase in the coating is small. On the contrary, the thermal stress is greater.

Figure 12a,b are macroscopic photos of the coating and substrate after 100 and 60 thermal shocks, respectively. It can be seen from Figure 12 that the first crack width of the coating is larger than that of the substrate. This may be related to the fact that the fracture toughness of the $La_2Zr_2O_7$ coating is lower than that of the GH4169 substrate. The lower the fracture toughness value of the $La_2Zr_2O_7$ coating, the faster the crack propagation, the larger the crack opening and the wider the crack width. On the contrary, the higher the fracture toughness value of the GH4169 substrate, the slower the crack propagation, the smaller the crack opening and the smaller the crack width.

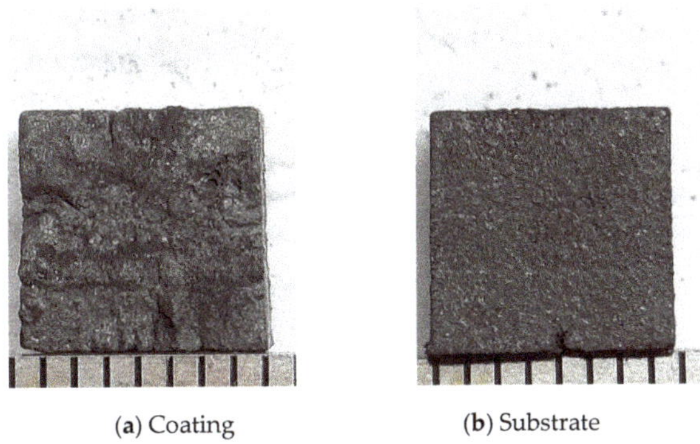

(a) Coating   (b) Substrate

Figure 12. Macroscopic photos of different samples after first crack thermal shock; (**a**) Coating; (**b**) Substrate.

It should be pointed out that if the laser-cladding method is used to prepare the thin LZ thermal barrier coating with a bond coat, the coating resistance to high temperature oxidation and thermal shock properties should be better than the coating without bond coat, but the former is more expensive to manufacture. In addition, the use of a bond coat is an essential step in the preparation of a thick LZ thermal barrier coating with excellent properties by laser-cladding. Finally, in order to obtain the thin LZ coating without cracks and pores, the laser-cladding parameters in this paper need to be further optimized.

4. Conclusions

In order to reduce the difficulty and cost of manufacturing and improve the high temperature oxidation and thermal shock properties of nickel-based superalloy, a thin $La_2Zr_2O_7$ coating without bond coat was successfully prepared on GH4169 substrate by laser-cladding based on pure $La_2Zr_2O_7$ powder. The thin $La_2Zr_2O_7$ coating is mainly composed of the $La_2Zr_2O_7$ phase. The high temperature oxidation property of the thin $La_2Zr_2O_7$ coating at 1100 °C for 110 h and the thermal shock property at 25~1100 °C are superior to that of the GH4169 substrate. The difference of the thermal expansion coefficient between $La_2Zr_2O_7$ and GH4169 has an important effect on the high temperature oxidation and thermal shock properties of the thin $La_2Zr_2O_7$ coating. The preparation method of the thin $La_2Zr_2O_7$ coating without the bond coat provided in this paper is suitable for a situation where the requirement of the coating lifetime is not long, such as the thermal barrier coating for launching satellite and missile warhead, which can reduce the manufacturing difficulty and cost. In addition, in order to obtain the thin $La_2Zr_2O_7$ coating without cracks and pores, the laser-cladding parameters in this paper need to be further optimized.

Author Contributions: Investigation—K.H., W.L., K.P., X.L., A.W.; writing—original draft preparation, W.L.; writing—review and editing, K.H. All authors have read and agreed to the published version of the manuscript.

Funding: This research received no external funding.

Acknowledgments: This work was supported by the Open Project Program (No. KF20180201) of Key Laboratory of Low Dimensional Materials and Application Technology, Ministry of Education, Xiangtan University, and the Fund (Grant No. SKLSP201913) of the State Key Laboratory of Solidification Processing in NWPU. The authors are also grateful to the Analytical and Testing Center of Huazhong University of Science and Technology.

Conflicts of Interest: The authors declare no conflict of interest.

References

1. Thakur, A.; Gangopadhyay, S. State-of-the-art in surface integrity in machining of nickel-based superalloys. *Int. J. Mach. Tool. Manu.* **2016**, *100*, 25–54. [CrossRef]
2. Wang, T.F.; Di, X.J.; Li, C.N.; Wang, J.M.; Wang, D.P. Effect of δ phase on microstructure and hardness of heat-affected zone in TIG-welded GH4169 superalloy. *Acta Metall. Sinica* **2019**, *32*, 1041–1052. [CrossRef]
3. Levi, C.G. Emerging materials and processes for thermal barrier systems. *Curr. Opin. Solid State Mater. Sci.* **2004**, *8*, 77–91. [CrossRef]
4. Cao, X.Q.; Vassen, R.; Stoever, D. Ceramic materials for thermal barrier coatings. *J. Eur. Ceram. Soc.* **2004**, *24*, 1–10. [CrossRef]
5. Doleker, K.M.; Ozgurluk, Y.; Karaoglanli, A.C. Isothermal oxidation and thermal cyclic behaviors of YSZ and double-layered YSZ/La$_2$Zr$_2$O$_7$ thermal barrier coatings (TBCs). *Surf. Coat. Technol.* **2018**, *351*, 78–88. [CrossRef]
6. Satpathy, R.K.; Mishra, B.R.; Mallick, B.; Mishra, B.K. Synthesis and application of nano-structured bi-layer YSZ-LZ thermal barrier coating. *Defence Sci. J.* **2019**, *69*, 185–194. [CrossRef]
7. Wang, Y.; Wang, L.; Liu, S.Y.; Liu, Y.; Wang, C.H.; Zou, Z.W. Nanostructured La$_2$Zr$_2$O$_7$(LZ)/8YSZ double ceramic layer thermal barrier coatings fabricated by thermal spraying. *China Surf. Eng.* **2016**, *29*, 16–24. (In Chinese)
8. Bobzin, K.N.; Bagcivan, N.; Brogelmann, T.; Yildirim, B. Influence of temperature on phase stability and thermal conductivity of single- and double-ceramic-layer EB-PVD TBC top coats consisting of 7YSZ, Gd$_2$Zr$_2$O$_7$ and La$_2$Zr$_2$O$_7$. *Surf. Coat. Technol.* **2013**, *237*, 56–64. [CrossRef]
9. Liu, L.T.; Li, Z.X.; Zong, Y.Y.; Hu, Z.; Li, J.T. Influence of laser treatment on the microstructure and properties of YSZ thermal barrier coatings. *Rare Metal Mater. Eng.* **2018**, *47*, 1238–1242. (In Chinese)
10. Xu, S.Q.; Zhu, C.; Zhang, Y. Effects of laser remelting and oxidation on NiCrAlY/8Y$_2$O$_3$-ZrO$_2$ thermal barrier coatings. *J. Therm. Spray Technol.* **2018**, *27*, 412–422. [CrossRef]
11. Soleimanipour, Z.; Baghshahi, S.; Shoja-razavi, R.; Salehi, M. Hot corrosion behavior of Al$_2$O$_3$ laser clad plasma barrier coatings sprayed YSZ thermal barrier coatings. *Ceram. Int.* **2016**, *42*, 17698–17705. [CrossRef]
12. Zhou, S.F.; Lei, J.B.; Xiong, Z.; Guo, J.B.; Gu, Z.J.; Dai, X.Q.; Yan, C.; Pan, H.B. Characteristics and properties of cryomilling-induced columnar growth in NiCrAlY coatings on Ni-based superalloy by laser induction hybrid cladding. *J. Mater. Res.* **2016**, *31*, 1338–1347. [CrossRef]
13. Renna, G.; Leo, P.; Cerri, E.; Zanon, G.P. Thermal shock behavior of CoCrAlTaY coatings on a Ni-base superalloy. *Metall. Ital.* **2015**, *7–8*, 33–41.
14. Pereira, J.C.; Zambrano, J.C.; Tobar, M.J.; Yanez, A.; Amigo, V. High temperature oxidation behavior of laser cladding MCrAlY coatings on austenitic stainless steel. *Surf. Coat. Technol.* **2015**, *270*, 243–248. [CrossRef]
15. Wang, H.Y.; Zuo, D.W.; Chen, X.F.; Yu, S.X.; Gu, Y.Z. Microstructure and oxidation behaviors of nano-particles strengthened NiCoCrAlY cladded coatings on superalloys. *Chin. J. Mech. Eng.* **2010**, *23*, 297–304. [CrossRef]
16. Chen, G.F.; Feng, Z.C.; Liang, Y. Formation mechanism of laser-clad gradient thermal barrier coatings. *T. Nonferr. Metal. Soc.* **2000**, *10*, 92–93.
17. Soleimanipour, Z.; Baghshahi, S.; Shoja-razavi, R. Improving the thermal shock resistance of thermal barrier coatings through formation of an in situ YSZ/Al$_2$O$_3$ composite via laser cladding. *J. Mater. Eng. Perform.* **2017**, *26*, 1890–1899. [CrossRef]
18. Zhou, S.F.; Dai, X.Q.; Xiong, Z.; Zhang, T.Y. Functionally graded YSZ/NiCrAlY coating prepared by laser induction hybrid rapid cladding. *Chin. J. Laser.* **2013**, *40*, 0403004-1–0403004-6. (In Chinese)
19. Ouyang, J.H.; Nowotny, S.; Richter, A.; Beyer, E. Characterization of laser clad yttria partially-stabilized ZrO$_2$ ceramic layers on steel 16MnCr5. *Surf. Coat. Technol.* **2001**, *137*, 12–20. [CrossRef]
20. Vandehaar, E.; Malian, P.A.; Baldwin, M. Laser cladding of thermal barrier coatings. *Surf. Eng.* **1988**, *4*, 159–172. [CrossRef]

21. Zheng, H.Z.; Li, B.T.; Tan, Y.; Li, G.F.; Shu, X.Y.; Peng, P. Derivative effect of laser cladding on interface stability of YSZ@Ni coating on GH4169 alloy: An experimental and theoretical study. *Appl. Surf. Sci.* **2018**, *427*, 1105–1113. [CrossRef]

22. Ouyang, J.H.; Nowotny, S.; Richter, A.; Beyer, E. Laser cladding of yttria partially stabilized ZrO_2 (YPSZ) ceramic coatings on aluminum alloys. *Ceram. Int.* **2001**, *27*, 15–24. [CrossRef]

23. Pei, Y.T.; Ouyang, J.H.; Lei, T.C.; Zhou, Y. Laser clad ZrO_2-Y_2O_3 ceramic Ni-base alloy composite coatings. *Ceram. Int.* **1995**, *21*, 131–136. [CrossRef]

24. Karaoglanli, A.C.; Doleker, K.M.; Ozgurluk, Y. Interface failure behavior of yttria stabilized zirconia (YSZ), $La_2Zr_2O_7$, $Gd_2Zr_2O_7$, YSZ/$La_2Zr_2O_7$ and YSZ/$Gd_2Zr_2O_7$ thermal barrier coatings (TBCs) in thermal cyclic exposure. *Mater. Charact.* **2020**, *159*, 110072. [CrossRef]

25. Wang, R.; Dong, T.S.; Di, Y.L.; Wang, H.D.; Li, G.L.; Liu, L. High temperature oxidation resistance and thermal growth oxides formation and growth mechanism of double-layer thermal barrier coatings. *J. Alloys Compd.* **2019**, *798*, 773–783. [CrossRef]

26. Jesuraj, S.A.; Kuppusami, P.; Kumar, S.A.; Panda, P.; Udaiyappan, S. Investigation on the effect of deposition temperature on structural and nanomechanical properties of electron beam evaporated lanthanum zirconate coatings. *Mater. Chem. Phys.* **2019**, *236*, 121789. [CrossRef]

27. Huang, B.Z. Research on high temperature properties of laser remelting plasma sprayed ZrO_2-8%Y_2O_3 thermal barrier coatings on GH4169 alloy. Master's Thesis, Nanjing University of Aeronautics and Astronautics, Nanjing, China, 2017. (In Chinese).

28. Zhou, H.F.; Yi, D.Q.; Yu, Z.M.; Xiao, L.R. Preparation and thermophysical properties of CeO_2 doped $La_2Zr_2O_7$ ceramic for thermal barrier coatings. *J. Alloys Compd.* **2006**, *438*, 217–221. [CrossRef]

29. Zhang, L.P.; Goncalves, A.A.S.; Jaroniec, M. Identification of preferentially exposed crystal facets by X-ray diffraction. *RSC Adv.* **2020**, *10*, 5585–5589. [CrossRef]

30. Guo, X.Y.; Wu, L.M.; Zhang, Y.; Jung, Y.G.; Li, L.; Knapp, J.; Zhang, J. First principles study of nanoscale mechanism of oxygen adsorption on lanthanum zirconate surfaces. *Physica E* **2016**, *83*, 36–40. [CrossRef]

31. Qu, H.X.; Kou, S.Z.; Pu, Y.L.; Guo, X.F. Effect of cooling rate on microstructure and mechanical properties of as-cast GH4169 alloy. *Foundry Technol.* **2016**, *37*, 481–484. (In Chinese)

32. Labrincha, J.A.; Frade, J.R.; Marques, F.M.B. Protonic conduction in $La_2Zr_2O_7$-based pyrochlore materials. *Solid State Ionics.* **1997**, *99*, 33–40. [CrossRef]

33. Zhang, J.; Guo, X.Y.; Jung, Y.G.; Li, L.; Knapp, J. Lanthanum zirconate based thermal barrier coatings: A review. *Surf. Coat. Technol.* **2017**, *323*, 18–29. [CrossRef]

34. Yu, Z.F.; Liu, L.; Liu, R.; Cao, M.; Fan, L.; Li, Y.; Geng, S.J.; Wang, F.H. Corrosion behavior of GH4169 alloy under alternating oxidation at 900 °C and solution immersion. *Materials* **2019**, *12*, 1503. [CrossRef] [PubMed]

35. Wang, D.X.; Tian, Z.J.; Shen, L.D.; Huang, Y.H. Research progress in preparation technology of thick thermal barrier coating. *Mater. Prot.* **2013**, *46*, 60–63. (In Chinese)

36. Wagner, C. Beitrag zur theorie des anlaufvorgangs. *Z. Phys. Chem.* **1933**, *21*, 25–41. (In German) [CrossRef]

37. Liu, M.T. Preparation and properties of $MoSi_2$-CoNiCrAlY nano-composite coating on GH4169 alloy by plasma spraying. Master's Thesis, South China University of Technology, Guangzhou, China, 2013; pp. 63–64. (In Chinese).

38. Li, M.S. *High Temperature Corrosion of Metals*; Metallurgical Industry Press: Beijing, China, 2001; p. 263. (In Chinese)

 © 2020 by the authors. Licensee MDPI, Basel, Switzerland. This article is an open access article distributed under the terms and conditions of the Creative Commons Attribution (CC BY) license (http://creativecommons.org/licenses/by/4.0/).

MDPI

St. Alban-Anlage 66

4052 Basel

Switzerland

Tel. +41 61 683 77 34

Fax +41 61 302 89 18

www.mdpi.com

Coatings Editorial Office

E-mail: coatings@mdpi.com

www.mdpi.com/journal/coatings

www.ingramcontent.com/pod-product-compliance
Lightning Source LLC
LaVergne TN
LVHW070908170726
843515LV00005B/730